DECISION MAKING IN
EMERGENCY MANAGEMENT

DECISION MAKING IN EMERGENCY MANAGEMENT

JAN GLARUM

CARL ADRIANOPOLI

Butterworth-Heinemann
An imprint of Elsevier

Butterworth-Heinemann is an imprint of Elsevier
The Boulevard, Langford Lane, Kidlington, Oxford OX5 1GB, United Kingdom
50 Hampshire Street, 5th Floor, Cambridge, MA 02139, United States

Notices
Knowledge and best practice in this field are constantly changing. As new research and experience broaden our understanding, changes in research methods, professional practices, or medical treatment may become necessary.

Practitioners and researchers must always rely on their own experience and knowledge in evaluating and using any information, methods, compounds, or experiments described herein. In using such information or methods they should be mindful of their own safety and the safety of others, including parties for whom they have a professional responsibility.

To the fullest extent of the law, neither the Publisher nor the authors, contributors, or editors, assume any liability for any injury and/or damage to persons or property as a matter of products liability, negligence or otherwise, or from any use or operation of any methods, products, instructions, or ideas contained in the material herein.

Library of Congress Cataloging-in-Publication Data
A catalog record for this book is available from the Library of Congress

British Library Cataloguing-in-Publication Data
A catalogue record for this book is available from the British Library

ISBN: 978-0-12-815769-5

For information on all Butterworth-Heinemann publications
visit our website at https://www.elsevier.com/books-and-journals

Publisher: Cathleen Sether
Acquisition Editor: Alexandra Romano
Editorial Project Manager: Michael Lutz
Production Project Manager: Omer Mukthar
Cover Designer: Miles Hitchen

Typeset by SPi Global, India

Contents

4. Decision making in emergencies, disasters, and catastrophic events

5. Common mistakes in decision making during events

6. Money is not the problem or solution

Chapter 7 Does the National Incident Management System (NIMS) really work for major event management?

Chapter 8 Silos will get someone killed

9. None of us are as smart as all of us

10. Global approaches to disaster management

11. Additional case studies

Author Biography

Jan Glarum has over 35 years of experience in the fields of EMS, Fire, Law Enforcement, Hospital, Public Health, and Emergency Management, including response to federally declared disasters. His experience includes an extensive background in planning, training, education, and response at the local, county, regional, state, and federal government levels, including Department of Defense initiatives CONUS and OCONUS. In 1999 he became a founding member of Oregon's Disaster Medical Assistance Team (DMAT) and continues his association with the team.

He has coauthored several books including, *Hospital Emergency Response Teams* and *Healthcare Emergency Incident Management Operations Guide*. Additionally, he has written numerous articles on emergency and disaster planning and response. He serves as a subject matter expert and speaker on emergency management, disaster planning, and has led hospital emergency response team development for hazardous materials events. He has developed a number of Incident Command System courses for hospital personnel to create operationally competent Incident Management Team members.

He is a Department of Defense validated medical CBRNE Subject Matter Expert, HSEEP qualified, and an ICS 300-400 instructor. He is a founding member of the Department of Homeland Security (DHS) and has led new course development and provided instruction for hundreds of students from around the United States at the Center for Domestic Preparedness. Programs included Weapons of Mass Destruction Incident Action Plan Development, Pandemic Planning and Response, EMS Response to Weapons of Mass Destruction, and Hospital Emergency Response Training courses.

Dr. Adrianopoli is an experienced professional with over 26 years of responsibility in preparing for, managing, and responding to the public health and medical consequences of natural occurring and man-made disasters including the CBRNE uses of Weapons of Mass Destruction. For 21 years he was the Regional Administrator/Regional Emergency Coordinator for federal Region V of the Office of the Assistant Secretary for Preparedness and Response in the U.S. Department of Health and Human Service (in various previous locations in the federal government, including FEMA for 3 years). He managed, assisted, or served at the Chicago Heat Wave of 1993, the Great Midwest Floods of 1995, the WTC attacks in 1993 and at 911, at Hurricane Katrina and at many other natural disasters to include serious floods, forest fires, ice storms, tornadoes, and even an extended deployment massing and repositioning assets for a potential asteroid crash (which, fortunately, occurred safely at sea) and many National Special Security Events. He co-managed the national TOPOFF Exercise in 2002 and played in TOPOFF Three and many others including a number of earthquake exercises coordinated by the Central United States Earthquake Consortium. In his role as Regional Emergency Coordinator he assisted in the development of three Disaster Medical Assistance Teams and one Disaster Mortuary Team and had overall responsibility for, and assisted in the development (along with his staff) of 24 Metropolitan Medical Response Systems.

In 2006 he was chosen by HHS as Co-Projector Manager to locate all National Disaster Medical System and related assets that were then located in FEMA and to arrange to bring them back to the U.S. Department of Health and Human Services. For nearly a decade he served on the Cook County (Chicago) Medical Examiner Emergency Response Team and was trained in forensic examination of deadly disaster scenes. All during this period he wrote articles, one book, and a chapter in another, and gave many presentations in the various aspects of emergency management preparedness, response, and mitigation.

Preface

This emergency management decision making (EMDT) textbook was born out of more than three decades of participating in and observing a wide array of disasters. Helping to reduce victim/patient mortality to as close to zero as possible remains our chief priority. Infrastructure, fiscal loss, and similar issues are important but remain secondary. As time passed it became clear that decision making was at the heart of our activities, in fact it was our major activity, sometimes managing significant aspects of disasters, and sometimes working with others. These included the nation's major terrorist events (in 1993 and 2001), numerous floods, serious winter storms (including ice storms), hurricanes, some huge tornadoes, a few deadly heat events, serious auto-train accidents requiring Disaster Mortuary Assistance Team participation, and even deaths at a major bridge collapse. Participation has included many training and disasters exercises (New Madrid, all the TOP Off exercises, the Y2K deployment), and even a predeployment, staging of assets for a potential hit by a significant meteor hit (avoided a populated area landfall), and massive predeployments of staged assets in case there were difficulties at some major professional sports events, many G-8, G-20, and similar defensive predeployments. The authors have travelled across the country many times to accomplish this all, and occasionally have gone out of the country to Japan, Canada, and many other lands. In addition to participation in a wide variety of emergency management situations, we have taught, researched, written books and articles, and presented papers in this subject area.

As we progressed writing this text we saw a two-part approach to developing this EMDT textbook that Elsevier wanted us to produce. First, we reviewed, summarized, and presented the study of decision making starting with Confucius, Greco-Roman philosophers through today's now popularized concerns for improving decision making. This included the century old Military Decision-Making Process (MDMP) a process that must be done correctly, and quicker than the enemy, or unnecessary deaths can result. Second, we saw that we could be most effective in reviewing the many situations that emergency managers will be called upon to make quick and accurate decisions, for example, in floods, hurricanes, and so on, and even terrorist events, where the results are often very similar to what occurs in naturally occurring disasters, injured patients needing treatment and transportation as quickly as possible, damaged infrastructure, even monetary issues regarding such complex concerns as flood insurance, FEMA reimbursements, and related complex issues. And of course there are decisions that must be made carefully within the public information and media areas as well as in the political world, outside of the agency structure, but often within it as well. Politics is a word that can have many definitions, and most of them have to be master by the wise emergency mangers at his or her peril.

We needed to weave all that we have learned, studied, and taught about decision making emergency management in disaster situations into a narrative that provided a perspective that was not the "normal or expected" way to address emergency management. This required adhering to a few basic rules for our findings. The first of which is that most serious emergency management errors occur at the higher levels of the command chain. While we can have high levels of assurance that the paramedic in the field or the emergency room physician or nurse will usually make correct lifesaving decisions, the same cannot be said with the same higher level of assurance for the upper command levels, especially if they are politically appointed. We also had to recognize that decision making at this level is difficult and subject to so many, often severe limitations.

All of this led us to view our task as one that required addressing many, often uncomfortable truths. We used obvious mistakes in decision making as well as obvious successes in a wide variety of situations, always with the intent of providing the coldest, most accurate take on situations we are capable of. Most often we deleted names when categorizing "bad calls" unless we dealt with public figures whose successes and failures have already been widely covered in the mass media. If we are successful with this text book, emergency managers reading it will begin to view their decision making as an important process that can be improved, but only with a strong investment of time and attention. In other words, an emergency manager cannot attain high levels of response and preparedness skills just by practice, knowledge, and even talent; effective, self-aware decision making is also required.

1

Introduction to decision making for emergency managers in perspective

This book was written by two, long time Emergency Managers to improve emergency management decision making in order to improve outcomes. To accomplish this mission the authors sometimes had to be coldly candid, discussing the underlying truths of many situations in which many believe there are no underlying truths, that what we see is mostly all there is. We will present some facts and situations that are rarely, if ever are openly addressed, but should be. In short, our many years in the field, going to deployments, writing books and articles, teaching, conducting or playing in exercises, and consulting give us the background to present emergency management decision making through many screens. Either or both of us have worked the first World Trade Center Bombing, in 1993, the 911 World Trade Center Bombing, the anthrax attacks that same year, the Great Midwestern Floods in 1995, the Chicago Heat Wave in 1995, and various Democratic or Republican Conventions, G-8s, G-20s, Hurricane Katrina in 2005 through Hurricane Sandy in 2013 and much of whatever happened in between these events (Fig. 1).

As one of the basic opening thoughts, the authors have observed that most serious organizational errors and problems come, in most instances, from the upper levels of the agency or organization, be they state, local,

1

FIG. 1   Making poor decisions at upper levels can have devastating impact to those in the field.

or federal. But for those Emergency Managers who are sitting atop their agencies or organizations, whether they are politically appointed or not, the complexities are even more immediate, and always threaten to crowd out the major missions and outcomes that are being sought. It should be no surprise that this level experiences more "mistakes or bad outcomes" than the paramedic, for example, who is conducting complex, but professional activities, with virtually no external constraints, political (partisan or organizational) or otherwise. Decisions at the higher levels can involve complex political decisions in situations for which the top manager has little authority. Also, the higher up one goes, the more likely is the fact that personal ego may play a stronger role in the decision-making process. And it is clear that people are more complicated than physical things and processes in virtually all instances.

For example, though a paramedic's opening an airway or starting an intravenous line can save a life, the decision process is quite rational and the options for action are relatively limited. Compare this to a FEMA Field Command Officer (FCO) who needs to call for the US Comfort, a Medical Support Ship that can bring hundreds of beds and medical staff in a few days, but who faces an outcry from local physicians who complain that their private medical practices soon will be wrecked by this massive provision of uncompensated (free) medical care. And there is a conservative governor whose staff agrees with the local doctors. There are no buses to bring the many patients to the USS Comfort when and if it arrives, though there is a sole contractor available who has a history of failed federal contracts with clear hints of incompetence. And the FEMA Regional Director is a powerful political actor of a different political party than the governor, who tends to be a bit of a micromanager where the FCOs are concerned (Fig. 2).

These last decades have seen a flurry of scientific interest in the fields of behavioral economics, psychology, business, and political science related to decision making. There has been strong mass media coverage focusing on the power of the "Gut" or instinctive aspects of decision making, particularly in areas in which we have knowledge and experience. Attention has also focused on the many cognitive biases, emotional and extraneous conditions that frequently degrade our otherwise "logical" decision making. These recent findings and insights will be summarized and applied to emergency management along with lessons from what the authors have experienced and learned in their cumulative decades in emergency preparedness, response, training, teaching, and research. The authors hope to convey that improving our decision making through "smart" procedures,

FIG. 2 Decision making can benefit from adopting methods to ease the effort. Much like using a roller system to move patients through a decontamination corridor.

knowledge, and strong efforts is still a difficult task, but one well worth mastering at any level. FEMA lessons learned, courses and administrative processes will be referenced when helpful but have, as we will discuss, often presented impediments to effective emergency management decision making because of the frequent and often disruptive changes in funding, restructurings, and priorities.

An attempt has been made to write somewhere between everyday language and disaster jargon, so that students as well as those familiar with emergency management can absorb the materials and avoid being bored. This will include a background consideration of community stakeholders, both as individuals and as representatives of governmental and corporate interests, along with the various linguistic, racial, economic, and cultural groups that may be affected by local disasters. This outside perspective seems opposed to the "can do," often judgment-based perspective of Emergency Managers and responders. The discussions will borrow from various fields to describe and analyze decision making from outside the emergency management "stovepipe." These findings and insights will be applied to help sharpen the decision-making process of Emergency Managers in everyday situations, as well as to gain deeper insights into the decision-making processes of those they must deal contend with, inside as well as and outside of the emergency management field. The fact that much of the research done in each field overlaps only testifies to the relevance and usefulness of what has been and continues to be learned across the world. The intent of the following pages is not to make an Emergency Manager an expert in psychology or any of the other fields that will be borrowed from, but to introduce them to these fields and to get them in the habit of looking at decision making as an important process they must master. To this end, the authors will include many references to assist both the practitioners of emergency management and students of the field, with the wide exposure to sources both within and outside of the emergency management field as well as to an inclusion of many examples from around the world. We know there is no worldwide emergency management and mutual aid system, but we are moving faster toward that end then most will ever notice.

And finally, before a discussion of some barriers to effective decision making as well as some potential guides to avoiding these constraints (when that is possible), it needs to be recognized that the complexity, the circumstances, and the procedures that surround emergency management decision making vary considerably, depending on the physical and organizational location of the decision making. As always, where a decision maker sits will have a strong effect on where a decision maker stands on an issue, or a problem.

If we are successful the book shelves of Emergency Managers having read this book won't just contain the latest Federal Emergency Management

Agency policies, some literature on various Weapons of Mass Destruction, floods, hurricanes, and other natural disasters and a few, worn copies of International City/County Management Association "green books" their second editions or similar background books. It will also eventually contain such books as Daniel Kahneman's *Thinking Fast and Slow*; Alan Jacobs' *How to Think: A Survival Guide for a World at Odds*; Sidney Finkelstein, Jo Whitehead, and Andrew Campbell's *Think Again: Why Good Leaders Make Bad Decisions and How to Keep It From Happening to You*; and Atul Gawande's *The Checklist Manifesto: How to Get Things Right* and perhaps even an older classic such as Herbert Simon's *Administrative Behavior: A Study of Decision-Making in Administrative Organizations*. This implies that to derive the most from these outside perspectives, the Emergency Manager must step outside his/her comfort zone and accept that while they may know disasters and emergency situations very well, others studying decision making have valuable lessons to share. Col. David Boyd, the brilliant Military analyst, whose decision-making theories anchor much of chapter two believed that to improve individual decision making, individuals have to develop and use as many different models of thinking as they can absorb. The authors have followed Col. Boyd's wise advice.

Chapter 2 opens with a discussion of wisdom and insights from over one century's thought and experience in military decision making, a body of knowledge too extensive and useful to ignore. Some of the key themes in the overall study of decision making, such as the value and the pitfalls of emotionally based, instinctive decision making, group decision making, potential problems with extremely high status or highly intelligent individuals, and personal biases in decision making will be covered more than once, in different perspectives. The constant theme of this book is that the Emergency Manager should be aware of his or her decision-making process, the difficulties in improving it, but the real benefits to improved outcomes of doing that. Without improved outcomes, there can be no improved decision-making process. The benefits of the improved decision-making processes should be in the improved benefits for the impacted/survivor populations.

In the following few paragraphs, some of the other major themes of this book will be addressed in summary, by using FEMA's failures during Hurricane Katrina response and recovery.

Some examples of FEMA staffing failures during the Katrina Hurricane period

We'll start with a brief Hurricane Katrina case study demonstrating how difficult it can be for high officials, especially those just appointed to their new roles or positions to avoid overconfidence leading to poor decisions. Our Katrina case study of failure by federal, state, and local decision makers begins when an experienced Regional Director (SES level) was contacted by then FEMA Director Michael Brown to immediately go to New Orleans and serve as FEMA's key manager on the ground. (Out of a concern for fairness, and in light of the complex and low staffing/information spot this FEMA manager was put into, we see no benefits in being name specific.) When the FEMA Regional Manager arrived, he immediately was bombarded with hundreds of emails a day, most from the White House. The Regional Manager was sent with no additional staffing, and later complained that the email burden alone was crowding out time to respond to the actual events they were witnessing, but not influencing to the extent that they otherwise could have. This Regional Manager had a long and distinguished career in FEMA as well as at the State emergency management level. Their reputations for knowledge were earned and well deserved and they were, no doubt, a wise selection by FEMA Director Michael Brown. However, despite having a huge and experienced regional office staff, none were requested to assist, save one GS 15, for a few days' service (Fig. 3).[221]

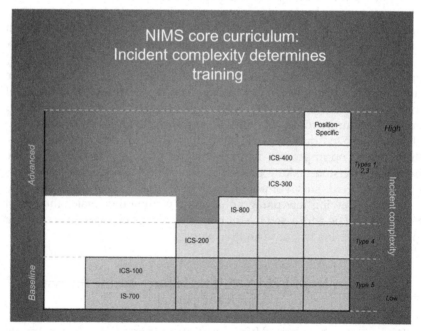

FIG. 3　Jurisdictions and agencies at risk from NIMS Type 1, 2, and 3 events, regardless if they are large or small require not just operational competency, but management competency as well.

Another senior manager was hired in the Katrina period to manage logistical missions. Again, this manager had a long and distinguished career, but was not yet "up to speed" in the existence and the location of many key, FEMA assets. It is quite possible that the "original sin" of widespread, short staffing at the senior and middle levels left FEMA in a position that its responses to Katrina would have been severely limited no matter what these recently hired senior officials did. Poor local and state responses only compounded the damage to the development of timely and well-resourced missions.

Under normal conditions, a high-level manager would have been well staffed, well briefed, and well resourced for this important and high-status assignment. Unfortunately, neither Michael Brown, nor Joseph Albaugh, the FEMA Director who preceded Michael Brown, were skilled or experienced in emergency management, though Albaugh had high level political experience, never a bad thing for an upper level manager to have acquired. Understaffing and resourcing were not just apparent in Brown's selection of a Regional Director to manage FEMA assets "on the ground" but in Brown and Albaugh's overall FEMA tenures. For example, after most of the high quality "Southern Senior managers" and others selected by James Lee Witt, President Clinton's FEMA Director, left their appointments during the year after Witt left FEMA, there is evidence that many SES, Headquarters Directorate slots remained unfilled. Research by Adamski et al.[5] found that after FEMA was placed in the new Department of Homeland Security (in 2003) that many response indicators had significantly deteriorated. The 17 FEMA Headquarters Directorates (However defined) were allowed to remain severely understaffed, limiting FEMA itself to compensate for weak state and local emergency management experience and staffing. But, if you don't have sufficient experience and knowledge in emergency management to understand the crucial value of SES and GS 15 leaders staffing the Directorates, and yet are confident in your own success (After all, Michael Brown and James Albaugh, in this instance, have been appointed FEMA Director) a disconnect may have appeared that was not going to be effectively addressed until personnel were changed.

Hindsight is an often cruel and inaccurate tool, if not used in a thoughtful and empathetic manner. But the Emergency Manager, at whatever level he or she is occupying has but two major requirements: (1) Gather information to analyze, define, and prioritize the missions, this includes reading relevant Government Accounting Office (or other) reports as well as FEMA's own, usually excellent, After-Action Reports, if they are available and (2) act to accomplish those missions. Just viewing the federal, mainly FEMA perspective of the Katrina missions, these were not done. The evacuation buses and other ground transportation, boats, rotary wing, the secured shelter facilities, medical care, and the massive supplies necessary were not provided, coordinated, and requested in a reasonable time perspective by FEMA, state or local authorities, but were, in fact often addressed by external leaders and organizations such as mutual aid (formally requested or not).

In fairness, an emergency management decision maker can have a huge and complex problem thrust at him or her almost "out of nowhere," can have little knowledge of what should be known about the problem and what others actually know and are doing about the problems. And if these constraints were not enough, the Emergency Manager may have little time to ponder the problems presented, save for an hour or two on a plane, without the ability to communicate effectively, ask questions, and receive information and advice. (Hopefully, though far from assured, from those qualified and experienced to do so effectively.) Overconfidence, as we will constantly stress, especially at the Senior Executive Service (SES) and G.S. 15 levels, can easily be a major enemy of effective decision making. For most SES and GS 15s, the routes to their grades have been long and hard. For the most part they've avoided most career-ending or damaging mistakes, they've had good mentors, usually worked diligently and justifiably feel proud of their accomplishments. And they are mostly pretty good at what they do in their areas of expertise.

Research findings by US Marine Scholars Stallard and Sanger (2014) can be applied to some aspects of both FEMA Directors (Albaugh and Brown) and even to the FEMA Regional decision making in not "Screaming for help," as soon as the catastrophic nature of Hurricane Katrina and his lack of adequate staffing assistance became evident. Stallard and Sanger point out that top managers, particularly those newly appointed to their positions, often fail at good decision making because of a "lack of humility" related to their own successes in attaining their own leadership positions. Hubris can also enable them to disregard wise advice by those charged with doing so. Their research found:

- Success can inflate a leader's belief in his ability to manipulate or control outcomes
- Success often leads to unrestrained control of organizational resources and,
- Success often leads to privileged access to information, people, and objects

In the first case study, unfortunately demonstrating these failures of decision making:

- Success can allow leaders to become complacent and lose strategic focus, diverting attention to things other than the management of their organizations.[6]

This introduction has demonstrated the many weaknesses with the rational choice theory in its many forms. As Herbert Simon has stated decades ago, there is no practical way for an individual to have all the knowledge about a potential decision.

2

The military decision making process

We begin coverage of the overall decision-making process with a review of the Military Decision-Making Process (MDMP) because the military has developed models of decision making under stressful if not deadly conditions, under which information is difficult to quickly gather, understand, and use. Faced with these limitations, use of the MDMPs provides structures that can focus and improve decision making in both warfare and serious disaster situations. MDMPs have, for centuries, been "field tested," researched, and improved to work under stressful and limited information conditions to solve field problems quickly and comprehensively (Fig. 1).

This is particularly appropriate to the field of Emergency Management where decision making is also often done under high stress, low-information conditions that allow little time for thoughtful consideration of a mission and the alternative methods of meeting mission requirements. To this extent the procedures that surround the MDMP can be particularly helpful in avoiding the many cognitive biases, such

FIG. 1   Military mission underway.

as unwarranted overconfidence that can hinder effective decision making. For example, the highly structured decision-making environment of a bank teller, a widow employee at a fast food outlet, or even a state driver's license inspector, all conduct their activities under tightly structured circumstances where human biases and inappropriate emotional reactions and frailties, and the lack of information necessary for making good decisions have been tightly controlled. The MDMP certainly cannot reproduce such structured circumstances but provides educated and trained structures to minimize human decision-making frailties, while maximizing human decision-making strengths and potential strengths. Some of the language and background materials used here are military and sometimes "clipped" but have been presented with much of the exact verbiage to protect the overall meanings.

The MDMP is a continuously iterative planning methodology to understand the situation and the mission in order to quickly develop a course of action, much in point for emergency management decision making. The MDMP is generally a 7-step process that is constantly being reevaluated and retailored to changing circumstances. Recall that it is oriented to the military and the battlefield. It starts with (1) Receive Mission, (2) Analyze Mission, (3) Develop Concept of Operations (COA), (4) Analyze COA (through War Games), (5) Refine the COA, (6) Approval Final COA, and (7) Produce Battlefield Orders and Their Dissemination.[7] MDMP systems have been designed to provide a uniform, widely shared structure to overcome individual weakness in decision making (e.g., from stress, emotions, inherent cognitive biases, incomplete information, changing circumstances, etc.) to assist in the making of quick and wise decisions leading to effective problem solving. The formal MDMP is applicable not only to the military but to the Homeland Defense Cooperative Agency (worldwide security through international partnerships) and to the field of Disaster Response.[8] The Lightning Press has developed a series of "Smart Books" in MDMP, many of which touch on areas such as leadership and related areas relevant to emergency management.

There have been recent attempts to incorporate up-to-date instinctive decision making into the MDMP (covered in this Chapter later), Col. William Boyd among them. Boyd made perhaps the wisest, and well-thought-out improvements of the MDMP in what he called the OODA Loop (Observe, Orient, Decide, Act) (Fig. 2). Boyd has been described by some as the "greatest military strategist that no one knows." Boyd served in the Second World War, in Korea and in Viet Nam as a much-awarded fighter pilot. At age 31, in 1961 he wrote "Aerial Attack Study," codifying the best dog fighting tactics for the first time, becoming the bible of air combat, revolutionizing the methods of every air force in the world. His Energy-Maneuverability (EM) Theory helped give birth to the F-15, F-16, and A-10 aircraft, working to improve his OODA Loop until the day he died in 1997. Some of the reasons that Boyd's findings had not been incorporated earlier into the wider MDMP were his unfortunate tendency to push back hard at anyone, regardless of rank or venue, who disagreed with or questioned his findings. Moreover, during his long involvement in the Military Reform Movement of the late 1970s and early 1980s he openly attacked the military bureaucracy with the support of Congress, leaving an open wound with the Air Force and Navy, limiting the acceptance of his other ideas, regardless of merit.[9] This was also unfortunate since modern research was recognizing the value of "Trained and educated" intuition, as summarized, for example, in Malcolm Gladwell's *Blink*.[10] Gladwell who observed, closely tracking Boyd's thoughts: "The

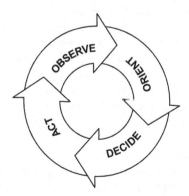

FIG. 2   Simple OODA loop.

very best and most successful...organizations of any kind are the ones that understand how to combine rational analysis and instinctive judgement," And, further:

> *What was the magical thing? It's wisdom that someone acquires after a life-time of learning and watching and doing. Its judgment.... what all the stories and studies and arguments add up to...is an attempt to understand this magical and mysterious thing called judgment...Judgment matters: It is what separates winners from losers...The key to good decision making is not knowledge. It is understanding. We are swimming in the former. We are desperately lacking in the later.*[10]

A representation of the simplified OODA Loop appears later, with a more detailed representation following. Notice the feedback loops and the detailed explanations that accompany the more complex version of the basic OODA Loop. The basic OODA Loop is an explicit representation of the process that all human beings and organizations use to learn, grow, and thrive in a rapidly changing environment—be it in war, business, emergency management, or in life, but Boyd had a much grander vision for it; it was to be the explicit representations of the always evolving, open-ended, far from equilibrium process of self-organization, emergence, and expanding mental perspectives (Ref. 11, p. 5). It was far from simple.

The Simple O.O.D.A Loop

As Boyd studied and expanded his OODA Loop concept, he made reference to Göedel's Incompleteness Theorems, Heisenberg's Uncertainty Principle, the 2nd Law of Thermodynamics, and even some explanatory theories of natural selection. We needn't be more specific, but these are mentioned to reference the depth of Boyd's thought. Ambiguity is central to Boyd's vision and not something to be feared, but to recognize that we never have complete and perfect information. (This theme will also be introduced by Herbert Simon, in this Chapter later.) Ambiguity and uncertainty surround us. While the randomness of the outside world plays a huge role in that uncertainty, Boyd argued that our inability to properly make sense of our changing reality is the bigger hindrance. When circumstances change, we often do not shift perspective and instead continue to try and see the world as we feel it should be. We need to shift what Boyd calls our existing "mental concept" in order to deal with the new reality. The crux of Boyd's case for why uncertainty abounds is that individuals and organizations often look inward and apply familiar mental models that have worked well in the past to try and solve new problems;

when the old models don't work, they will often keep using them and trying to make them work. Charlie Munger (Associate of Warren Buffett at Berkshire Hathaway) calls this tendency to stay with the familiar even in the face of change the "man with a hammer syndrome." From the old saying "to the man with only a hammer, everything is a nail." So it is with folks with few mental models to work with; every problem can't be solved with their current thinking, so they keep hammering away, confused and disillusioned their work isn't producing expected results (Bret and Kate McKay, pp. 3, 4):

> It is a state of mind, a learning of the oneness of things, an appreciation for fundamental insights known in Eastern philosophy and religions as simply the Way [or Tao]. The Way is not an end but a process, a journey....(with) the connections, the insights that flow from examining the world in different ways, from different perspectives, from routines examining the opposite proposition, were what were important. The key is mental agility.[12]

A more complex diagram of the OODA Loop

Observation

A discussion of the detailed OODA Loop follows and starts with the first term of the OODA Loop, "Observation." Some of this is quite complex and may take a few rereadings but should be worth the effort. To observe, from a tactical standpoint, to effectively observe, you need to have good situation awareness. For example, if you are a security professional, start keying in on where all exits are whenever you enter a public building, how rooms and floors are connected, and begin to visualize how an armed intruder would be confronted, in each location. This is especially important as recent Homeland Directives recommend directly and quickly confronting a shooter, robbing his initiative, upsetting his plans and expectations by having him adjust to an attacker, as opposed to fleeing victims posing no threats to his plans or self. By observing, taking into account new information about changing environments, our minds become open systems rather than closed ones. We gain the knowledge and understanding that's critical in forming new mental models; placing ourselves in Condition *Yellow*, best described as *relaxed alert*. As an open system, we're now positioned to overcome confusion-inducing mental entropy (Fig. 3).

In his presentation of the OODA Loop Boyd noted that we'll encounter two problems in the Observation Phase. (1) We often observe imperfect information and (2) we don't necessarily understand the information we are seeing.[13]

These two pitfalls are solved by developing our judgment—our practical wisdom. Even if one has perfect information it is of no value if it is

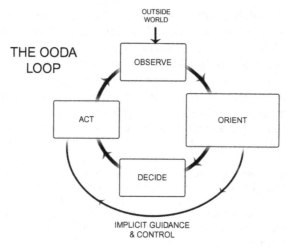

FIG. 3   More complex OODA loop.

not coupled to a penetrating understanding of its meaning, if one does not see the patterns. Judgment is key. "Without judgment, data meant nothing. It is not necessarily the one with the more information will come out victorious, it is the one with better judgment, who is better at discerning patterns." This theme of developing and using our instinctive powers will be much more fully developed later.

> Observe" means more than just "see," it's something more like "actively absorb the entire situation." Observation includes your own situation...and the environment more broadly; It includes all the dimensions of that environment: the physical, mental, and moral dimensions. The observation phase is data gathering in the broadest possible sense.... You are not just looking at your own numbers on the screen you are looking at the emotional context, industry trends ...Imagine you were a perceptive financial trader that understood the OODA Loop in the run-up to the 2008 financial collapse. In the observation phase, you saw that the market was on its way towards record-highs. You felt the mental dimension. Many people felt the market could only go up. You saw there was a huge increase in financial instruments including mortgage-backed derivatives. You saw that many...people who were taking out mortgages had...lower incomes than people taking out mortgages... earlier.[14]

Orientation

Orientation is the critical OODA Loop step, but it is most often overlooked (Ref. 11, p. 7); it includes understanding one's heritage and previous experiences, then analyzing and synthesizing that with the observations you've made. The goal is to find mismatches: errors in your or others' previous judgments. A general rule: bad news is the best kind as long as you catch it in time when it can be turned to your advantage (Ref. 14, p. 11). Orientation is where our mental models exist, and it is our mental models that shape how everything in the OODA Loop works. Osinga[13] writes, "Orientation shapes the way we interact with the environment...it shapes the way we observe, the way we decide, the way we act. In this sense, orientation shapes the character of present OODA Loops, while the present loop shapes the character of future orientation" (Jim and Ret McKay, 2014, p. 7 citing Orsinga). Two factors that affect your OODA Loop during Orientation are Denial and Emotional Filter. Denial is when you refuse to accept or deny that this is happening to you. Emotional Filter is a similar to Denial except that it emphasizes the emotional aspect, as in "Oh man, please don't let this happen to me." Both of these response can and will affect reaction time but fortunately they can be overcome with training.[15] Also see https://www.facebook.com/DoctrineMan/ a military management and decision-making site that continually features such scholars and practical experts as Col. Boyd and those who still follow and use his brilliant analysis.

So how do you orient yourself in rapidly changing circumstances? You constantly have to break apart your old paradigms and put the resulting pieces back together to create a new perspective that better matches your current reality. Boyd has called this "creative destruction," and when we do this we analyze and pull apart our mental concepts into discrete parts. Once we have constituted these basic elements, we can start what Boyd calls "creative induction," using these old fragments to form new mental concepts that more closely align with what we have observed is really happening around us (Ref. 11, p. 7, 8). Lubitz et al.[16] (p. 571) stress, speaking directly to Emergency Managers, that Boyd's Orientation stage is when actors begin to reassert his/her control over the environment into a cohesive, predisaster configuration. This is nothing less than the act of "getting one's bearings" in the postdisaster chaos by groupings of the disorganized structure of the disaster environment into a cohesive reality of easily recognizable blocks, and then aligning blocks into even larger and better organized mental assemblies (cognitive maps of the disaster environment). Providing buses at the Superdome during Katrina for the purpose of evacuating survivors from a now dangerous, filthy, and overcrowded place instead of evacuating survivors to the Superdome, was an example of changing a response based on a changed cognitive map. The lack of adequate security was one of the factors that changed the cognitive map of the Superdome though the news stories about violence, rape, and mass lawlessness were overblown.

Again, citing Charlie Munger, stressing the value of a varied library of mental knowledge and models: "...you've got to have models in your head. And you've got to array your experiences... vicarious and direct— on this lattice of models...the first rule is that you've got to have multiple models—because if you just have one or two that you're using, the nature of human psychology is such that you'll torture reality so that it fits your models, or at least you'll think it does…**So you've got to have multiple models. And the models must come from multiple disciplines—because all the wisdom of the world is not to be found in one little academic department** (Bold emphasis by the McKays). Boyd believed and acted on this principle, as he began investigating theories seemingly as far afield as natural selection as another model that explained parts of human behavior (but he showed wisdom and courage in doing so as military experts do not seem, at first blush, as individuals who have much of a stake in evolutionary biological principles). Boyd left military thinkers (and Emergency Managers) with helpful ideas about the value of learning multiple theories or models of human behavior Boyd fostered (Ref. 9, p. 23 in Ref. 11, pp. 8–12).

As an important part of his Orientation mode, Boyd recommended that the more mental models you have at your disposal the more you have to work with in creating new, more useful ones. Boyd warned his audience of the way in which strict operational doctrines can stifle cultivation of a robust tool of mental models. He noted that the Army had its doctrine, the Navy had its doctrine, and the Air Force had its doctrine. He felt doctrines have a tendency to harden into dogmas, and dogmas have the tendency to create folks with "man with a hammer syndrome." He said "…read my work, 'doctrine does not appear in their once. You can't find it. You know why I don't have it in there? Because it's doctrine on one day, and every day after it becomes dogma…" It was for this reason that Boyd advocated for familiarizing yourself with as many theories and fields of knowledge as possible, and continuing to challenge your beliefs, even when you think you have them figured out.

Boyd and Munger gave suggestions of thought models a wise follower of the OODA Loop would study: Boyd's list included Mathematical, Logic, Physics, Thermodynamics, Biology, Psychology, Anthropology, Conflict (Game Theory). Munger's surprisingly similar list includes: Math Accounting (and its limits), Engineering, Economics, Probability, Psychology (specifically the cognitive biases that cause us to make terrible decisions—covered in detail in the next section of this Chapter), Chemistry, Evolutionary Biology (can provide insights into economics, history, and statistics). Boyd stressed the value of destroying and creating mental models, maybe even starting a journal covering concepts in the various new mental models you have begun to learn. He also stressed that the wise follower of the OODA Loop needs to constantly be orienting his/her ideas, because the world around you is constantly changing, and the models that best explain and even predict it will have to be changing too, if they are to remain relevant. And finally, before an operation (a disaster situation), Boyd stresses that you want to be fairly confident that your mental models or concepts will work before you actually need to use them. To do this you train, exercise externally, but internally you study what mental concepts have and haven't worked in similar situations and then practice, train, and visualize using those mental concepts. For example, if TV messages before and right after a hurricane have worked poorly in the past, communications plans to alert citizens regarding evacuation routes, shelters, and related items, perhaps it is time to add newer social media methods to augment the older, never fully satisfactory models. Perhaps reaching disadvantaged, minority citizens through their churches, or other citizens through organizations such as the American Legion of the Veterans of Foreign Wars should now be major, not minor parts of an effective communications plan.

Decide (hypothesis)

Boyd didn't spend much time articulating the Decide step except that it's the component in which actors decide among the best alternatives generated in the Orientation phase. For Boyd it is impossible to select a perfect matching mental model because we often have imperfect models to use, so we're forced to settle for ones that aren't perfect, but good enough. Finally, Boyd felt that decision making was essentially moving forward with our best hypothesis, our best educated guess about which mental model will work (Ref. 11, p. 12). Contrary to this, decision making will receive extensive coverage as its own process, below. But there is a logical reason for this. For example, a fast food restaurant clerk or an accountant often make cold, accurate decisions where an individual's cognitive biases, emotions perspectives, and related potential decision-making difficulties are not apparent and certainly not needed. The wider MDMP and the OODA Loop follow like strategies in attempting to orient the decision process to one totally directed at the mission, thereby avoiding potential human foibles, with one major exception—when intuition (alternately called judgment or even wisdom) has been developed by experience and knowledge to work quickly in discerning patterns and finding appropriate solutions to utilize to the maximum extent possible.

It is here where Boyd's primary feeling to have the warrior (and certainly the Emergency Manager) constantly develop new mental models, new information, and new experiences in order to meld perfectly with the types of knowledgeable and experienced intuition that Malcolm Gladwell praised in *Blink: The Power of Thinking Without Thinking* and the researchers into thinking and decision making he was summarizing. Gladwell wrote of the powerful, instinctive ability of the mind, without conscious thought, to review huge amounts of existing information to discern patterns and make quick, instinctive decision "in a blink," the so called "gut" or "instinctive" based decision making. But Gladwell also wrote that these decisions made in a blink can only be useful when they are accompanied by thousands of hours of work and thought in a specific area. A chess player, a great surgeon, even a highly skilled Emergency Manager, when viewing situations well within their various fields of expertise, are examples of individuals who can likely make accurate decisions about what to do next, in a "blink." Unfortunately, when individuals have not prepared their minds with thousands of hours of practiced actions in a specific area, the gut or instinct-based decisions they may have such confidence in are usually wrong, because they are based on nothing but their own biases, desires, and choices. These are the uninformed types of decisions that Boyd (and Gladwell) are attempting to train our minds to avoid. And being the fallible creatures that we are, those who are highly skilled and trained in one decision-making area too often assume that their highly skilled and trained decision-making process, often done in a "blink" will work in different areas. It usually will not.

Action

The fourth and final step of the OODA Loop is putting the COA into action, testing it, receiving feedback and testing it again as a continuing process. We should all be constantly experimenting and gaining new data that improves how we operate in every facet of our lives. As Osinga notes in *Science, Strategy and War*, actions "feed back into the systems as validity checks on the correctness and adequacy of the existing orientation patterns." Action is how we encounter it if our mental models are correct, for example in the next serious flood in New Orleans, if levees in the Ninth Ward breach, are people still on roofs after a few hours, not having been rescued, and directed to shelters without adequate police support. Summarizing his OODA Loop, Boyd wrote:

> We gotta get an image or picture in our head, which we call orientation, Then we have to make a decision as to what we're going to do, and then implement the decision...Then we look at the (resulting) action, plus our observations, and we drag in new data, new orientation, new decision, new action, ad infinitum....[11]

Boyd almost seems to have been writing specifically for emergency managers whose must make quick decisions based on imperfect information in rapidly changing circumstances where their previous thought models may no longer be relevant. In a direct application of the OODA Loop to emergency management, von Lubitz et al.[16] (p. 567) wrote, as they stressed their categorization of Boyd's OODA Loop as the development, testing, retesting and application of "Actionable Knowledge":

Every disaster introduces a dramatic change in the affected environment. The informational content is massively increased by a number of new, often poorly understood elements (decreased environmental transparency). The orderly nature of original information that the environment contained and by which it was characterized prior to the disaster (granularity) is not disrupted, and the granularity of the environment increases...However, new information continuously generated during the entire time course of the critical events obscures situational awareness and impairs disaster-mitigation efforts. *Actionable knowledge* derived through the process of effective, real-time management and fusion of new, disaster-generated information with equally, efficient use of pre-existing knowledge provides the essential tool with which to increase transparency and reduce the granularity of the disaster environment (improving...the ability to respond to sudden and unpredictable challenges that the disaster environment may generate....

In addition to the MDMP items that we have discussed, there is another military "thought and action asset" that all Emergency Managers need access to. For decades the US Department of Defense has supported the development of the kinds of information and data required to develop enhanced situation awareness ("Ground truth" as most Emergency Managers refer to it) all across the disaster spectrum across the nation. To accomplish this mission 5 or 6 highly skilled and credentialed officers, serving as Joint Regional Medical Planning Officers (JRMPOs) are continually deployed across the nation, also serving at significant disasters and NSSEs. Each JRMPO carries a highly detailed, constantly updated electronic file (called a "Smart Book") addressing regional demographics, public health, and medical assets across the nation, and a complete listing of Department of Defense medical assets, many of which are available to support disaster preparedness, response, recovery, and mitigation missions. The wise local, state or federal Emergency Manager knows his or her JRMP and is never much farther than a text message or a call away. The JRMPOs are key support to FEMA Regional Offices, members of the Regional Advisory Council (RAC) attend monthly meetings of the Regional Interagency Support Committees (RISC), assist in training and exercise missions and are frequently not only called upon by ESF-8 but by virtually all of the ESFs. The Defense Coordinating Officers (DCO) are also assigned regionally and are members of the FEMA RAC, and though they have the widest wealth of Department of Defense assets, they do not travel as much as the JRMPOs and generally operate at a higher organizational level across all of the ESFs, as opposed the JRMPOs who focus on ESF#8.

As a final observation on the MDMP. There is a Facebook website on military decision making which may be the most read and popular of the MDMP sites. For those interested in this rich and constantly relevant issue, the following Doctrine Man website is attached (https://www.facebook.com/DoctrineMan/). Also, the extensive War Room websites address

MDMP topics and a wide variety of other related topics, among them many interfaces with antiterrorism, especially of interest to Emergency Managers who have strong concerns in these areas. The website can be accessed through warroomeditors@gmail.com.

A short history of the study of decision making

Before moving to more modern findings on decision making, it is valuable to put decision making in a historical perspective. Leigh Buchanan and Andrew O'Connell writing in the 2006 issue of the *Harvard Business Review* conduct an exhaustive review of thoughts on decision making since prehistory.[17] For millennia human decision making had been guided by interpretations of entrails, smoke, dreams, and the like. Even the Greeks consulted the Oracle of Delphi and various seers, who continued to have strong influences through the Middle Ages; though even earlier Confucius said that decisions should be informed by benevolence, ritual, reciprocity, and filial piety, quite an improvement it would seem. Plato described the interplay between emotions and reason as two horses pulling in opposite direction, a view that has persisted in some form or another, to this day. Remarkably, Aristotle foreshadowed the intensive future research covering reason, emotions, external physical circumstances, and internal biases affecting decision making. He was seeking shaped and informed thinking and decision making to develop individuals who would live virtuous lives. Of course, Aristotle was not discussing effective thinking in response to disasters. But as we will see, he could have been. What we need, he stated, in order to live well, is proper appreciation of ways in which such goods as friendship, pleasure, virtue, honor, and wealth fit together as a whole:

> In order to apply that general understanding to particular cases, we must acquire, through proper upbringing and habits, the ability to see, on each occasion, which course of action is best supported by reason. Therefore practical wisdom, as he conceives it, cannot be acquired solely by learning general rules. We must also acquire, through practice, those deliberative, emotional, and social skills that enable us to put our general understanding of well-being into practice in ways that are suitable to each occasion.[19]

Darwin would argue that the influence of emotions on decision making has survived the rigors of natural selection so must have some intrinsic merit. Darwin's thoughts are valid because emotions give us useful guidance whenever the environment we are in fails to provide us with all the information needed for thoughtful analysis and can also make us act quickly and decisively when necessary.[19] It was not until the Renaissance introduced the beginnings of a science-based approach to the world, in the West at least, that theories on decision making start to substitute facts for fears and myths. More modern theories of decision making began with the classical economic assumptions that people make generally rational choices about scarce resources, based on their own self-interests. Since at least the mid-1950s this has been referred to as the rational choice model and is the most widely known decision making model—based as it is, on an idealized, self-interested and nonemotional "Economic Man," who, in the act of being "economic" is also "rational." By the 1970s there were two broadly accepted ideas about human nature. First, people are generally rational, and their thinking is normally sound. Second, emotions such as fear, affection, and hatred explain most of the occasions on which people depart from rationality. But just as many political scientists were almost universal in their accepting the economically derived, rational choice model, and using it as the basis of thousands of research studies, Herbert Simon, and soon, many others began challenging the validity of the rational choice model as being unrealistic if not actually irrational. He stated:

> The social sciences suffer from acute schizophrenia in their treatment of rationality. At one extreme, economists attribute to economic man a preposterous omniscient rationality. Economic man has a complete and consistent system of preferences that allow him to always choose among the alternatives open to him; he is always completely aware of what those alternatives are; there are no limits on the complexity of the computations he can perform in order to determine which alternatives are best; probability calculations are neither frightening nor mysterious to him.[4]

Another early critic of the rational choice model,[21] found the model limited, and frequently wrong, and that often decision makers are altruistic, and not just totally self-centered rationalists. She argued convincingly for a more complex interpretation of behavioral motivations that also included duty, honor, public spirit, respect, and love.

Incrementalism

While the rational choice model was being almost universally rejected, some fields, particularly urban planning, a field closely allied to emergency management, were finding some aspects of rational choice worth saving. What was being proposed was not to accept the "rational" definition of the decision maker, but to salvage at least some of the "perfectly rational process" as worth reviewing and holding as an ideal, albeit one that was not attainable. The "old" and mostly rejected rational model follows, using emergency management as the field using the model:

1. *Define the problem* (many injured, many homeless, damaged housing, communications, transportation, mass power outages, health and medical systems severely damaged).
2. *Clarify values* (e.g., Fairness, how important is it to open shelters and roads for the homeless poor, clear roadways using local contractor who may be slower to organize than external contractors; is it less expensive and effective to bring patients to newly opened facilities as opposed to reopening existing but severely damaged hospitals).
3. *Select goals.* Having gone through Steps 1 and 2, we are now in a position to choose one or more goals relative to the problems initially defined by the interplay of both data/information and values.
4. *Formulate* alternative response plans or programs.
5. *Forecast the consequences of the alternatives developed in the previous step.*
6. *Evaluate and select one or more courses of action (alternatives).*
7. *Develop detailed response plans.*
8. *Review and evaluate.* Once response or recovery has begun it is necessary periodically to review the process and results to date with a view to deciding whether the original response and recovery should be modified or left as developed.[22]

Critics of the "ideal" rational decision process contend that few problems are so easily categorized so that they can be easily be solved by using the logical process described. They also contend that value clarification sounds easy, but that that it often cannot be done because there is so little unanimity regarding commonly accepted values. A final point critics make is that practical matters, such as time, cost complexity, or inability to reach agreement on values and goals are practical problems not likely to be solved as decisions are made.[22]

Responding to the challenge of a clearly unrealistic, decision making process based on rational choice, as well as the fact that the decision makers themselves were often not rational in their decision making process, political scientists Charles E. Lindblom and David Braybrooke decided to try and find a way between outright rejection of all that the rational decision making model stood for, and saving some aspects of the model, especially those that left it in place as an ideal, albeit one that could never be attained. They summarize the problem:

> It seemed plausible to suggest that what economists, other social scientists, policy analysts, and decision-makers generally do in the face of complex problems, even when they try to be rational, does not at all approximate rational decision making. The clue to how they normally do achieve defensible analyses of their policy problems seemed to be in further development of the incremental concept-leading to an account of analytic practices that focus on alternatives differing only incrementally, in a political system that normally offers only that range of alternatives.[23]

Looking even further into actual practices in evaluating and deciding on alternative public policies, Lindblom found that they succeed, where conventionally conceived decision making does not, in taking intelligent account of the cost in time, energy, and other resources—of analysis, as well as of the impossibility, for those sufficiently complex problems, of bringing an analysis period to an end. In other words, he suggested start searching for the most effective decision, by searching for those that have already been decided, and making only incremental changes in those. He called this process "disjointed incrementalism." They based this theory on their contention that in the actual practice of public decision making, there are existing choices, they referred to as "adaptations" that were used by decision makers, which then conventional rational choice theories could not account for. In other words, public decision makers were holding rational choice up as their model process, but not following it at all. Instead they would go back to previous decisions in an area that worked and make small changes to the process as times and circumstances changed[23] (Fig. 1).

Needless to say, this process of disjointed incrementalism is what the wise Emergency Manager does when faced with an allocation of resources (regarding timing, amount, cost, etc.) should first talk to others and look at existing after-action reports, news accounts, and so on, of how the resources issue was addressed in previous disasters. It is valuable to review disjointed incrementalism, not just because of its inherent insights and wisdom, but also because urban planners still study the concept, and urban planning is a key foundation of FEMA thinking. Critics of the incremental decision making model observe that the model cannot work on new problems because there are no past decisions to view and change

FIG. 1   Complex incidents will test any emergency manger's ability to make decisions.

incrementally. Critics of the approach might also argue that excessive reliance on the incremental approach can make a decision maker excessively dependent on precedent and past experience and thus blind to worthwhile new ideas or to avoid old mistakes. This means that excessive reliance on incrementalism can lead to excessive caution and missed opportunities.[22]

Even as Allison was a realistic counterpoint to the too often unrealistic rational thinking model, sociologist Amitai Etzioni suggested a "mixed scanning" model that fit between each. The idea was simple, Etzioni advocates a two-step decision process. The first step being a general scanning process that is conducted to get an overall picture and to decide which elements merit more detailed examination and analysis. He used the example of weather monitoring systems using space satellites and stressed that the mixed scanning might have a few parts, when a large field is scanned down to a smaller one, and perhaps even a smaller one again. Etzioni argue that his model avoids the excessive commitment to precedent and past experience inherent in the incremental model, while avoiding the assumptions of "complete information" that the rational model almost assumes. In fact, this mixed scanning approach mirrors the practical approach that many Emergency Managers take when faced with the huge array of facts and players that must be addressed almost all at once in some large disasters. Some have criticized this mixed scanning approach by saying that it is little more than what "sensible and no dogmatic" planners or Emergency Managers would do anyway; most of whom cannot afford to take either the strict rations approach or the incremental approach and will necessarily use some synthesis of both. Etzioni simply formalized and made explicit what most actually do (Ref. 22, pp. 278–81 citing "Mixed scanning: A 'Third Approach to Decision Making,' based on Ref. 222), Andreas Faludi, ed. pp. 217–30). People running agencies share power, and often differ about what should be done, and the differences do matter.

In the 1990s critics of the rational choice model were being reinforced and supported by findings in social psychology, behavioral economics, and business/management that found that decision making was subject to many irrational faults, but there was strength, in many situations, in the instinctive, emotionally based aspect of decision making. Emergency Managers who claim to make many decisions not so much on facts but on their instincts, their "guts," may not have been all wrong. But, as it turns out, much more is involved.

Classical, rationally based economics was being quickly replaced by a behavioral economic model demonstrating the power of the pull of small amounts of evidence was such that even those who knew about it should resist being succumbed. "People's intuitive expectations are governed by a consistent misrepresentation of the world." By expanding this concept researchers found that intuitive human expectations were often governed by a consistently erroneous model of the world. In 2002, Kahneman and later, Thaler in 2017, received Nobel Prizes in economics and psychology borrowing from the two with their evolving concepts of decision making. Previously, instinctive thinking making had neither been fully understood nor valued enough in previous "scientific" approaches to decision making. And now the biases inherent in both intuitive System 1 thinking as well as more rational System 2 thinking were being recognized.

Thinking fast and slow in decision making

Most Emergency Managers make decisions quickly, often under stress, while facing serious distractions. They base their decisions on a combination of instincts and knowledge, with advice or criticism from others. Frequently, Emergency Managers' decisions are made without their paying much attention to the process of decision making itself, and of their own limitations and strengths. Fortunately, in recent years, decision making, particularly instinctive ("gut level" or "sixth sense") decision making, increasingly has been studied, written about, and discussed across this nation and much of the world. Popular and very useful books and articles have included (1) Malcolm Gladwell's *Blink* which stressed the value of quick, emotionally based, intuitive decision making, done in a "blink," as well as situations when this powerful tool fails spectacularly; (2) a whole issue of the *Harvard Business Review* (2006) that discussed decision making in the widest perspective; and (3) psychologist Daniel Kahneman's landmark work, *Thinking, Fast and Slow*, which compared and contrasted slower, more rational thinking, which he named System 2 thinking with faster, emotionally and instinctively based System 1 thinking; Kahneman and Tversky, two psychologists who, along with Thaler formed the newer field of behavioral economics (as opposed to the classical "rational economics" which was not really rational at all) also considered cognitive biases and predispositions that can hinder as well as facilitate intuitive decision making, and the "noise" that can cloud effective decision making. Decision making biases are systematic errors such as overoptimism or harmful stereotypes that can damage our decision making, but other biases and predispositions can also provide shorthand rules of thumb that can save time rarely available for thoughtful pondering of issues. Decision making noise, such as current mood, the time since the last meal, the weather, or even a toothache also can affect judgment in even the most serious of circumstances. These researchers along with extensive media coverage have emphasized how complex, often irrational, but none the less powerful instinctual capacities affect, often dominating, both rational and instinctive decision making. In short, we are rarely as effective at decision making as we believe (Fig. 2).

FIG. 2   Odd as it may sound, decision making regarding the dead is more straight forward that dealing with survivors.

According to Alan Jacobs,[27] System 1 provides us with a repertoire of useful biases and predispositions that reduce the decision making load on our conscious brains. These biases aren't infallible, but they provide helpful rules of thumb; they're right often enough that it makes sense to follow them and not try to override them without some good reason. Again, according to Alan Jacobs,[27] we simply would not be able to navigate through life without these biases, these prejudices—the cognitive demands of having to assess every single situation would be so great as to paralyze us. So we need the biases, the emotional predispositions, to relieve the cognitive load. We just want them to be the right ones. We pass through life mainly depending on Kahneman and Tversky's intuitive System 2; the more rational, System 1 kicks in only when we perceive a problem, inconsistencies, or anomalies that need to be addressed. Jonathan Haidt, another psychologist, uses different terms. He thinks of intuitive thinking as an elephant and conscious decision making as the rider. The point is that our intuitive decision thinking is immensely powerful and has a mind of its own but can be gently steered by the rider who is skillful and understands the elephant's inclinations. Or as Jonathan Haidt[28] summarized very succinctly, "Intuitions come first, strategic reasoning second…. or, the mind is divided, like a rider on an elephant and the rider's job is to serve the elephant." He concludes by stating his feeling that the elephant is 99% of our mental processes and goes on to demonstrate the strength and logic of his position.

A classic example of powerful and effective System 1 thinking was presented by psychologist Gary Klein[20] who tells the story of a team of firefighters that entered a house in which the kitchen was on fire. Soon after they started hosing down the kitchen the commander heard himself shout, "Let's get out of here!" without realizing why. The floor collapsed almost immediately after the firefighter's escape. Only after the fact did the commander realize that the fire had been unusually quiet, and his ears had been unusually hot. Together, these impressions prompted what he called a "sixth sense" of danger. He had no idea what was wrong, but he knew something was wrong. It turned out that the heart of the fire had not been in the kitchen but in the basement beneath where the men had stood. We have all heard such stories of expert intuition: the chess master who walks past a street game and announced, "White mates in three," without stopping.

Expert intuition strikes us as magical, with the typical example of a skilled physician just looking at a patient and diagnosing her almost immediately, but it is not magical. Each of us performs feats of intuitive expertise many times each day. Most of us have absolutely no difficulty in detecting anger in the first words of a telephone call or that a driver acting erratically in the lane next to ours is dangerous. Our everyday intuitive abilities are not less marvelous than the string of thoughts of an experienced firefighter or physician-only more common. Despite all of these constraints and limitations, Emergency Managers can never forget that "every success, every mishap, every opportunity seized or missed is the result of a decision that someone made or failed to make."[25] Before there is disaster preparedness, response, or recovery there is a set of decisions made, good or bad, wise or unwise, that once made are difficult, sometimes impossible to modify or rescind.

Effective instinctive decision making doesn't just appear. In many respects this capacity must be developed. Good intuitive judgment, for example, an art expert working at the Museum of Art in New York, taking one look at a fraudulent masterpiece and knowing at a glance that it is a forgery, developed his instinct by spending evening after evening taking things out of cases and putting them on a table. "There were thousands of things. I mean we were there every night until ten o'clock, and it wasn't just a routine glance. It was really poring and poring and poring over things." What he was building, in those nights in the storerooms, was a kind of database in his unconscious. This type of effort is expended by experts who have learned to recognize familiar elements in a new situation and to act in a manner that is appropriate to it.[10] But we don't have to be experts to have developed these strong instincts in certain circumstances, whenever we have something that we are good at—something we care about—experience and passion fundamentally change our first impressions. This does not mean that when we are outside our areas of passion and experience, our reactions are invariably wrong. It just means that they are shallow. They aren't grounded in real experiences. It would be like taking an experienced Emergency Manager and placing him in an art gallery and having that manager decide on whether or not a piece of art work was real or not, or in having that art expert sit in an Emergency Operations Center and begin making decisions of resource deployment with little objective information to guide her. T.S. Eliot reminds us, "When we do not know, or when we do not know enough, we tend always to substitute emotions for thoughts."[27] As Kahneman observes, "Unfortunately, professional's intuitions do not all arise from true expertise,"[20] almost like getting stock market advice from your physician or a cab driver.

In the past, most courses or trainings in emergency management and many other fields that focused on decision making included the admonition, "Honestly confronting your individual biases, and doing your best to work around them, is the first step to improved decision making." This old rule still applies, but what has been learned in recent years teaches us that this is much harder than once thought, because to a great extent, we are our biases.

Not surprisingly, improving individual decision making will be difficult. Thinking is hard, as Allan Jacobs reminds us, and many of our own instincts will try and pull us back to what is comfortable, if not effective. The same powerful, emotionally based intuition that can size up a situation in an instant can also present strong biases that rob other decisions of any semblance of common sense or accuracy. Timothy D. Wilson[30] writes in *Strangers to Ourselves*: "The mind operates most efficiently by relegating a good deal of high-level, sophisticated thinking to the conscious, just as a modern jetliner is able to fly on automatic pilot with little or no input from a human 'conscious' pilot. The adaptive unconscious does an excellent job of sizing up the world, warning people of danger, setting goals, and initiating action in a sophisticated and efficient manner." Humans have limited time and brainpower and as a result they use simple "rules of thumb" to help them make judgments. This is all further complicated, as Jonathan Haidt reminds us "One of the greatest truths in psychology is that the mind is divided into parts that sometimes conflict." He quotes the Roman poet Ovid, "I am dragged along by a strange new force. Desire and reason are pulling in different directions. I see the right way and approve it but follow the wrong.[227]

So much has been learned about decision making in recent years (though some streams date back to antiquity) that an investment here can pay off in better, wiser decision making for Emergency Managers during disasters as well in the other parts of their lives. Again, improving our decision making will not be easy, it requires an investment, but the outcomes are well worth that investment. The types of people that run toward disasters when most others are fleeing them are among those that will benefit most by improved decision making.

The emergency management decision making perspective here is action and outcomes, not policy oriented and it will encompass decision making across the field. To realistically address decision making in emergency management it will be necessary to adopt the widest agency and stakeholder perspective, starting at the higher levels of emergency management and stakeholder organizations, where political appointees or top, nonpolitical managers operate, as well as down the pyramid to the lower levels of organizational decision making. It will also be necessary to address the external conditions and individuals whose own decision making affects the Emergency Managers, to whom this book it directed.

These next sections expand on System 1 intuitive decision making ability and how it both helps and hinders System 2, rational decision making. Simply put, our brains are well suited for some tasks, but ill-suited for others. The consequences of this reality can be wide ranging: from simple illusions, to annoying memory glitches, to irrational decisions whose effects can just as likely be innocuous as fatal. We will discuss over and over the difficulties involved with overcoming cognitive biases, noise and situations leading to bad decisions and bad outcomes.

We will present some of the key findings to Emergency Managers as well as a number of experts and websites to help them get deeper into areas that interest them. The purpose is to have Emergency Managers and students view their own decision making with a bit more knowledge and insight, but also with a heightened skepticism about their own decision making and that of those they must work with.

Barriers to effective decision making and some methods to try to avoid them

The framing effect

Framing effects are different ways of presenting the same information to evoke different emotions, resulting in different decisions based on that same information. Saying "the odds of survival one month after surgery are 90%" is more reassuring than making an equivalent statement "mortality within one month of surgery is 10%." Similarly, describing cold cuts as "90% fat free" is more attractive than describing them as "10% fat." Jumping to conclusions, the wrong ones, is much more likely depending on how information is framed. Evidence that our brains can be so easily eluded also comes from situations where bad decisions can lead to horrible outcomes. For example, people's views about something as important as child custody cases can yield different outcomes depending on whether they are asked, "Which parent should have custody of the child?" or "Which parent should be denied custody of the child?" In this instance Parent A had a modestly good listing of attributes including good income, health, working hours, and a rapport with the child. Parent B had an above average income, a very close relationship with the child, an extremely active and work social life, a good deal of work-related travel, and minor health problems. When the questions were phrased regarding positive attributes, parent B's impressive credentials with regard to income and relationship with the child won over parent A's more modest abilities on these fronts. Ask who should be denied custody, however, and a very different picture surfaced. The strategy yielded evidence of parent B's inadequacies as a guardian regarding the busy social and work life and health issues.

The familiarity effect

Dozens of studies have demonstrated that merely having been exposed to something, whether it's a face, image, word, or sound, makes it more likely that people will later find it more appealing. This is the same familiarity bias that is exploited in advertising and that can also be exploited in politics. Telling the same facts or even distortions of facts over and over again, "The other party is against Medicare" or "....these tax cuts are actually for you and have worked over and over again in the past" seem to keep appearing in American politics and working again and again. When Nazi propagandist Goebbels stated that by telling the same emotionally oriented, group-centered lie repeatedly, it would eventually be accepted as truth by most people, he used what many researchers have since demonstrated.

This bias is also a relatively "hidden" source of concern because it can easily strike the professional who has built up a career long cache of "things and ideas" that remain firmly in the memory, for better or worse. While it is true that experience does help build up intuitive decision making in some key areas, it is also true that a history of bad habits, if not failures in an area can also have its effects, and separating these two types of experiences may not be that easy. The Emergency Manager must always fight this tendency by attempting to overcome these effects, but it is not easy. Having some skeptical management or staff colleagues willing to "tell the cold truth" to anyone can help as long as the truth is not wielded as a sharp object.

The confirmation bias

People have a tendency to seek answers and make decisions that tend to confirm their existing beliefs and positions on issues. When they seek additional information often they don't seek neutral information, but they

FIG. 3   Team work is essential in making good decisions.

tend to seek only confirmatory information. Two existing mental biases contribute to the Confirmation Bias: people overestimate the probabilities of unlikely events and this leads to them overweighting unlikely events in their decision making. The probability of a rare event will often (often, but not always) be overestimated, because of the confirmatory bias of memory. Thinking about that event, you try to make it truer in your mind. A rare event will be overweighed if it specifically attracts attention.

This presents some interesting situations for Emergency Managers, on one hand each successive emergency deployment builds a cache of circumstances, decisions, and outcomes that can clearly be judged and that tend to enhance his intuitive, fast thinking decision making. But when the Emergency Manager is challenged to pull out certain aspects of that response from memory, the tendency will be to assume they were good response memories, and to assume they are still valid as guidance. Of course, those memories may not have been examples of the best response. Perhaps a good solution to this problem is, again, to have individuals surrounding the Manager with a track record of successful response activities. And what had been noted more than a few times, it is usually easier to judge the value of other's decision making abilities, than one's own. In this respect a later section, on Group Think, addresses some of the weaknesses as well as the real strengths of that group approach (Fig. 3).

The halo effect

If you like a president's politics, you usually like his voice and his appearance as well. The tendency to like everything about a person, including things you have not observed, known as the "halo effect," has been known about for over a century, though knowledge of it has not reached the general public, and "This is a pity," as Kahneman[20] observes, because the "Halo Effect" is a good name for a common bias that plays a large role in shaping our view of people and situations. It is one of the ways that System 1 decision making generates a simpler and more coherent view than the real thing. Presumably this works in the opposite direction with people you don't like, and all of whose views you will tend to dismiss. To avoid this common error in decision making, Kahneman recommends deriving useful information from multiple sources of evidence, always trying to make these sources independent of each other as possible.

> The principle of independent judgements...has immediate applications for the conduct of meetings, and activity which executives in organizations spend a great deal of their working days. A simple rule can help: before an issue is discussed, all members of the committee should be asked to write a very brief summary of their positions. This procedure makes good use of the value of diversity of knowledge and opinion in the group. The standard practice of open discussion gives too much weight to the opinions of those who speak early and assertively, causing others to line up behind them.[20]

For the Emergency Manager who depends on a few, tried and true staff members for information sources, it would appear best to keep these, but to gather a few more, perhaps farther away from the "trusted veterans," perhaps even from without the inner circle of Emergency Management.

Group think

Nearly 50 years ago, Irving L. Janis[44] changed how agencies and corporations thought about the effectiveness of decision making in groups. Unfortunately, too often those using the term "Groupthink" are not guided by what he found, leaping to the wrong conclusion that it was the nature of groups themselves, usually committees, to make bad decisions (e.g., the old joke that a camel is a horse designed by a committee). Janis wondered why powerful, otherwise knowledge leaders acting in small groups made such disastrous decisions (The Bay of Pigs, The Invasion of North Korea that started the Korean War, and the Vietnam War) when some were involved in such brilliant and long-standing decisions such as the development of the Marshall Plan. Janis sought the answer in the psychology of group dynamics. Of course, he knew that groups could be subject to distorted thinking by fear, anger, elation, even irrational prejudices. But Janis felt that these reasons for human failure, though accurate, were missing something. It was not enough to say the bad decisions were made because people are fallible or "To Err is Human." History is full of instances where group participation has brought forth noble instincts in people, we have to look no further than selfless citizens risking their lives to help others in the aftermath to Katrina, or any recent disaster. On the other hand, the powerful, unconscious actions of crowds have perpetrated massive evil from the terror after the French Revolution, the Nazi depredations, and the Holocaust to the roving bands of ethnic murders in the dissolution of Yugoslavia in the 1990s. An experienced Emergency Manager knows that serious disasters are accompanied by potentially destructive group activities (frequent looting and vandalism), as well as positive group activities including many examples of citizens placing themselves in danger to help others. Janis narrowed his focus to very powerful, small groups tending to make worse decisions in significant areas than they would if they were acting as individuals, away from the group.

Studies of industrial organizations indicate that while some groups fostered higher productivity and conscientiousness, some similar groups fostered slowdowns and socializing activities that reduced productivity. The same type of variations in groups' outputs may be found among policy-making groups in large organizations. Janis found that in studies of social clubs and other small, powerful groups, conformity pressures have frequently been observed. Members who voiced views contrary to those held by the group were eventually driven from the group or were ignored. This drive for uniformity of beliefs damaged decision making. Janis also found the tendency of groups to develop stereotyped images that dehumanized out-groups against whom they are engaged in competitive struggles, so that group discussions become polarized, sometimes shifting toward extreme conservatism and sometimes toward riskier courses of action than the individual members would otherwise be prepared to take.

In short, we can summarize Janis' findings, which fit nicely with more modern findings on decision making members of powerful, high status groups value their continued participation in such groups more highly than breaking with conformity, even when the decisions being made are disastrous. Does this mean the Emergency Manager must dispense with the small group decision making process that is reflected all across FEMA's National Incident Response System? Of course not, but it should serve to make Emergency Managers skeptical of their group decision making as well as the decision making in groups that will affect their roles. Using the wisdom imparted by what has been learned, it is always sensible to view a committee regarding members that are too "chummy" and cooperative, too dominated by one or two dominant voices, either too lacking in subject matter expertise or short of it, and all of the other potential pitfalls.

As a final perspective on group decision making, Cohen, March, and Olsen[40] give us the Garbage Theory and the surprising truths they uncovered. In their histories in academia they found a few keys to committee activities that seemed to hold up regardless of the type of committee, where the committee meetings were being held, and who was attending them. They called organizations "organized anarchies," that were subject to three decision making problems:

(1) Problematic preferences, which meant people did not define their preferences very clearly that much as political actors often fail to (or refuse to) define their goals. Yet as some argue, people act in the absence of clearly defined goals and action is often precipitated by fuzzing over what one is trying to accomplish.

(2) Unclear technology: Many committee members did not know organizational processes, their own jobs, and had only a modest knowledge of why they were doing what they were doing. They operate a lot by trial and error, by learning from experience, and by pragmatic invention in crisis.

(3) Fluid participation: Participants drift in and out of decision making, so the boundaries of such an organizations are more fluid than settled. The time and effort of individuals differ greatly, who shows up for or is invited to a given critical meeting, and their degree of activity at the meeting can make a huge difference. As Woody Allen observe, "Just showing up is 90% of success." Despite all of this, March, Cohen, and Olsen remind us that the participants made decisions, adapted, and survived, at least after a fashion and sometimes quite well.

In perspective, are these "organized anarchies" all that different from participants from FEMA, the Department of Defense, and the Emergency Support Functions, some on their first missions, some gnarled veterans, often attempting to balance field work with attendance at a state, local, or county emergency office, attend meetings and still be under the direction of their agency supervisors, with the FCO and possibly even the FEMA Regional Director added to the mix. And sometimes, when the standard 12-h shifts stretch out considerably longer during the first week or two, so that exhaustion and too many colds and coughs are added to the potential restraints on good decision making. And it all seems to work out fairly well, most of the time.

The true believers

Eric Hoffer[41] reminds us that those who have surrendered their beliefs to a mass movement or cause that promises to immediately and drastically improve the lives of the "true" believers as well as their belief in improving the world by adherence to the cause, which can be religiously, socially, or politically based, and can be very difficult individuals to sway with anything as small as logical arguments. Hoffer also reminds us that strict adherence to such movements, the doctrines they preach, and the programs they project also can breed fanaticism, strong enthusiasm, fervent hope, and even hatred and intolerance, for those not in the group. It is not likely that Emergency Managers will often encounter full-throated true believers espousing their cause while responding to a disaster. But it can happen. Nonetheless, each of us, those we interface within the Emergency Management System, as well as those external to it have elements of the True Believer. It is not unwise to classify strong 2nd Amendment advocates as well as strong gun regulation advocates, intensely political conservatives or liberals, or even intensely patriotic Americans as opposed to those who consider themselves to be more "world citizens" to have at least some aspects of the True Believer within themselves. Emergency Managers must contend with many shades of True Believer that may surface during the tense emotional circumstances that accompany each disaster they are called upon to address.

The "smart person" problem

Unfortunately, professional's intuition do not all arise from true expertise. As noted, a physician's effective and even impressive diagnostic abilities do not necessarily qualify him/her to give useful and expert opinions, say, on financial obligations and a condominium's due diligence regarding its narrow range of borrowing and investment choices. An individual who had developed professional intuition in the finance area would likely be a more appropriate guide. But after years of being "the smartest guy in the room," it is not surprising to see that confidence moves into areas where it is not warranted. And, as cognitive psychologists have found, most of us are already more optimistic about our decision making abilities than our experience and our knowledge supports. Tali Sharot[50] observes that when intelligent people are presented with accurate, factual challenges to their deeply held beliefs they just come up with even more clever rationalizations regarding why their beliefs are accurate. Jacobs[51] quotes Avery Pennarun, an engineer at Google, "Smart people have a problem, especially (although not only) when you put them in large groups. That problem is an ability to convincingly rationalize nearly anything....What I have learned, working here, that smart, successful people are cursed. The curse is confidence. It's confidence that comes from a lifetime of success after real success, an objectively great salary...." Of course intelligence is useful in disaster response, as the complexity of decision making and the power of those involved only grows, the further up the management chain an Emergency Manager ventures. But without knowledge or even insight into a particular area, unsupported, but confident "expertise" can be a severely limiting factor. And this weaknesses can affect us all.

The irony is that although people may or may not equate high intelligence with competence, they do judge competence by combining their perceptions of two dimensions of strength and trustworthiness. It should not be surprising that Hemant Kakkar and Niro Sivanathan writing in the *Harvard Business Review*[51] observe that during uncertain times, and disasters qualify, we prefer dominant leaders who are confident, controlling, and strongly hierarchical. Many of these traits are positive, but dominant leaders have also been known to exhibit negative transits such as narcissism, aggression, and uncooperativeness. They are prototypical "alpha male" in the group, and they frequently claim leadership positions instead of waiting to have leadership responsibility conferred upon them as they develop talent to deserve that leadership role.

Though no research of which we are aware confirms it, it appears likely that a good number of FEMA's top managers as well as top Emergency Managers at state and local level, powerful corporate stakeholders, and politicians at all levels are reflective of this intelligence "confidence trap." Some strategies for addressing this as well as other restraints on good decision making will be addressed later, but, as always, being aware of the issue and some of its consequences is at least a start in dealing effectively with some of the most intelligent or credentialed that are involved with decision making in disaster situations, "normal" as well as the black swan variety. The prestige pathway to leadership, on the other hand, is associated with individuals who are respected, admired, and held up as examples. As a counterbalance to these points, Roy Armstrong, a senior manager in the "old" US Department of Health, Education and Welfare, before it was truncated to the US Department of Health and Human Services made the wise observation (to one of the authors as a student intern) that it was much better to "fight a smart guy instead of a dumb guy." Pardon the vernacular of the day, no disrespect was meant to anyone. "Fighting the smart guy was easy, all you had to do was make a couple of smart, but illogical, even contradictory moves. The "smart" guy will be tied up for days reading so much into even your simplest bureaucratic moves. And the more false starts the more confused will the smart guy with your "illogical" moves. The dumber guy will just keep coming after you, directly, straight on until he kills you or you nail him."

Simple mathematical formulas often make better predictions than professionals

Kahneman demonstrated that what he called "algorithms" or simple mathematical formulas could predict some events better than trained psychologists or other professionals. Psychologist Paul Meehl initially reviewed the results of 20 studies that had analyzed whether *clinical predictions* (type of diagnosis, etc.) based on subjective impressions of trained professionals were more accurate than mathematical, *statistical predictions* made by combining a few scores or ratings, according to a simple rule. The simple formulas were more accurate in 11 of 14 studies. Eventually the number of studies he reviewed approached 200, which included the evaluation of bank credit risks, the length of hospital stays, the odds of recidivism of juvenile offenders, and even the future price of Bordeaux wine—all of which were predicted more accurately by small numerical rules than by the various professional experts in these areas, using their judgment.

After 30 years of similar research findings, Meehl,[52] commented, "There is no controversy in social science which shows such a large body of qualitatively diverse studies coming out so uniformly in the same direction as this one," simply stated, small numerical rules or formulas were more effective at many predictions than were professionals in their respective areas of expertise. The Princeton economist and wine lover, Psychologist Orley Ashenfelter has offered a compelling demonstration of the power of simple statistics to outdo world-renowned experts, in this instance to predict the price of wine. He used just three predictors, the temperature over the growing season, the amount of rain at harvest, and the total rain in the year before the harvest. The correlation was .90 between Ashenfelter's prediction of the cost per bottle and the actual price of the wines that were sold. All this research and more, in so many cases of judgments by auditors, pathologists, psychologists, organizational managers, and other professionals suggest that the normal level of inconsistency is typical, even when cases are reevaluated a few minutes later, most to the same conclusion. To maximize predictive accuracy, final decision should be left to mathematical formulas, especially in low-validity environments.

The Black Swan: The impact of the highly improbable

A Black Swan is a very improbable event with three characteristics: It is mostly unpredictable; it carries massive impacts; and, after the fact, we concoct an explanation that makes it appear less random and more easily predictable than it was. The astonishing success of Google was a black swan, so was the World Trade Center bombing. But why don't we acknowledge black swans until after they occur? According to Nassim Nicholas Taleb[224] part of the answer is that humans are hardwired to learn specifics when we should be focused on generalities. We focus on things we already know, and time and time again fail to take into consideration what we don't know. We are, therefore, too often unable to effectively and quickly estimate opportunities. Taleb has studied how we fool ourselves into thinking we know more than we actually do and restrict our thinking

FIG. 4   Decision makers need to recognize what they know as well as what they don't know about any given situation.

to the irrelevant and the inconsequential while large events continue to shape and reshape the world (Fig. 4).

Taleb was not directing his work to improving the responses and recoveries to the such huge and mostly unpredicted disasters as the Chicago's Deadly Extreme Heat Event of 1993, the World Trade Center Bombings, 2001, Hurricane Katrina in 2005, the Myanmar (Burma) Cyclone of 2008, and the Haiti Earthquake, 2010 but his descriptions of our failures as a system (and as key individuals) to realistically address black swans can assist in future catastrophic responses, to some extent. Taleb stresses that much of our general lack of ability regarding responding to and even learning from "black swan" events comes in part from what has been called the "narrative fallacy" or the System I's need to try and explain events and situations in logically and connected manners, even when there is no linkage between the events. Failure to deal adequately with black swan events also comes from what has been referred to as the confirmation bias, as described before, reflecting our tendency to use new facts and situations to confirm our existing feelings and predispositions, even when they may be in direct opposition. In fact, as Tali Sharot[42] reminds us, when you provide someone with new data, data that seems to contradicts their preconceived notions, can cause them to come up with new counterarguments that further strengthen their original view, this has been called the "boomerang effect." Again, changing our ideas is difficult, not impossible, but clearly difficult. All the Emergency Manager, or anyone for that matter can do is to try and get in the habit of recognizing the biases and preconditions that can lead us to respond poorly to black swan events, and to attempt to learn from them as the single events that they likely are. Of course, this assumes we are sensitive to the events we should be categorizing as these singular lack swan events. That in itself will take some training of our memories in those areas.

Thoughts on how to think well from Allan Jacobs

Unlike most authorities cited in this background chapter on decision making, Allan Jacobs is not a psychologist, behavioral economist, or business/management expert but a well published freelance writer with a strong reputation for analytic pieces. He accepts Kahneman and Tversky's description of our intuitive, System 1 minds and our rational thinking, System 2 minds, but is more hopeful regarding our ability to overcome systemic biases and other barriers to effective decision making, especially when we develop the habit of attempting to become aware of them and attempting to counteract them.

Jacobs introduces a few key concepts we can all recognize, concepts that are almost magnets for the kinds of biases and errors in thinking psychologists like Kahneman and Finkelstein, Whitehead and Campbell have stressed. He cites the strong human tendency to join and associate in groups and the tendency, over time, for the individual "members" to begin adopting the group's biases, stereotypical thinking, and even its negative views of others simply because they are not associated with the group.[27] This concept has gotten a lot of popular media coverage under the term "tribes," and describes a strong potential problem that can be reflected in political, ethnic, income, and all too many other groups. He stresses that though we may have begun associating with individual members because of characteristics we initially enjoyed in them, we may be unaware of other effects we might not have seen nor appreciated. Jacobs stresses that we begin to be influenced by the emotional dispositions of the group's members as soon as we "join." According to Jacobs our choice of group associations are critical to our mental status, and one of the less difficult (but certainly not easy) ways to protect ourselves from habits of thinking that are negative and that are not accurately reflective of the surrounding reality. Many of our emotional dispositions are not necessarily the results of our own hardwiring, but are reflective of what we have absorbed from individuals and circumstances external to us. In short, Jacobs stresses the importance of avoiding those individuals and groups we do not want to resemble because our associations with others do shape our unconscious, whether we notice it or not and whether we approved of the changes or not, most of which we will be unaware of, even as we adopt them. Or as you might have been warned by your mother to "avoid bad companions so you don't end up acting like them."

Jacobs critically reviewed much of the significant recent research on decision making as well as some much older ones and derived twelve "antidotes" to flawed decision making. He views thinking as an art, but a difficult one, one that is to be done wisely and ethically, much in accord with the ancient Greeks, as opposed to the scientifically, research oriented perspective presented by most other authors included here.[27]

- When faced with provocation to respond to what someone has said, give it five minutes. You have probably entered a "Refutation mode," and your response, made just moments later, may be wiser, less emotionally driven.
- Value learning over debating, don't "talk for victory."
- As best as you can, online and off, avoid the people who fan the flames.
- Remember that you don't have to respond to what everyone else is responding to in order to signal your virtue and right-mindedness.
- If you *do* have to respond to what everyone else is responding to in order to signal your virtue and right-mindedness, or else lose your status in your community, it is probably a community you would be better or leaving.
- Gravitate as best as you can, in every way you can, toward people who seem to value genuine community and can handle disagreement with equanimity.
- Seek out the best and fairest-minded of people whose views you disagree with. Listen to them for a time without responding. Whatever they say, *think it over*.
- Patiently, and as honestly as you can, assess your repugnancies.
- Sometimes the "ick factor" is telling; sometimes it's a distraction from what really matters.
- Beware of myths and metaphors (categorizations) of thoughts or groups of people that have meanings beneath the surface that are wrong-headed or unethical. And never forget that there is power in myths and metaphors that can be inaccurate and highly destructive.
- Try to describe others' positions in the language that they use, without indulging in the temptation to describe in words or phrases that convey negative definitions that you have developed to define the other group.
- Be brave.

A final "piece" of Alan Jacob's wisdom: We shouldn't expect morel heroism of ourselves. Such an expectation is fruitless and in the long run profoundly damaging. But we can expect to cultivate a more general disposition of skepticism about our own motives and generosity toward the motives of others. And—if the point isn't already clear—this disposition— is the royal road that carries us to the shining portal called Learning to Think (Ref. 27, p. 147).

And, in the spirit of the thoughts of Alan Jacobs, the authors believe that we as Emergency Managers need to be kind (which is not the same thing as being weak). Will a person that does not like a leader but that nonetheless respects her, inform her of a bureaucratic storm about to engulf her, or will the Emergency Manager just keep walking past her desk, content that "staying in my own lane," is always safest, and leave it at that.

Governmental effects and constraints on good decision making

It has often been observed about government staff that: "Where you stand on an issue depends on where you sit." This piece of conventional wisdom has more than a bit of truth to it; a lot of your views as an employee are shaped by the culture, the organizational and political needs, and even the policies of the agency that employs and pays you. This is mostly self-evident but was not stressed by those studying decision making. Well over 50 year ago political scientist Graham T. Allison was studying the Cuban Missile Crisis, which, at the time threatened to cause World War III, pitting the nuclear powers of the US and the USSR against each other under very dangerous circumstances. The Cuban missile crisis was a seminal event; for 13 days in October 1962, there was a higher probability that more human lives would end suddenly than ever before in history. Had the worst occurred, the death of 100 million Americans, over 100 million Russians, and millions of Europeans as well would make previous natural calamities and inhumanities appear insignificant. That such consequences could follow from the choices and actions of national governments obliges students of government to think hard about these problems. As he studied the Crisis Allison saw that the organizations that housed the decision makers affected the substance of the decisions they made, something that, until that time, was never consistently stressed by serious students of government. Allison proceeds from the premise that marked improvement in our understanding of such events depends critically on more self-consciousness about what observers bring to the analysis and what each analyst sees and judges to be important is a function not only of the evidence about what happened but also of the "conceptual lenses" through which he or she looks at the evidence and the assumptions and categories employed by the agency analysts as governmental actors (Ref. 47, pp. 248, 49). Written many decades before the psychological findings regarding both the fallibilities and the strengths of "thinking fast and thinking slow" of individual and group decision making Allison focuses on the strong effect agency culture has on decision making.

Shortly after Allison wrote, Niskanen[48,49] developed the theory that the main driving force of public administrators is to maximize their own budgets in order to enhance their salaries and power. Though that is no doubt correct in some instances, the criticism does not appear to be supported by an significant research or even common sense.[50] Public salaries at the top levels are miniscule in comparison to analogous salaries in the private sector.

Allison saw three different conceptual models that explained the wise (and a few foolish) decisions made by the US and the USSR in avoiding turning the missile Crisis into the Third World War. These three decision making models have shaped the perception of governmental (agency) decision making ever since. The purpose of Allison's study was to explore some of the fundamental assumptions and categories employed by agency analysts in thinking about problems of governmental behavior. These models have been used by political scientists for decades in helping them to understand and even predict some facets of an agency's decision making processes. They are quite applicable to FEMA, DHS, USEPA, US Army Corps, and all of the federal, state, and local agencies that are involved in emergency management decision making. Never forget, agency staff act in certain ways because for the situations, their own responses and talents, and because of the constraints placed on them by the agencies for which they work. An Emergency Manager who fails to at least consider the effect that different agencies have on emergency management decision making is not a very realistic Emergency Manager despite whatever levels of talents and networks that individual may have. Some of the following can get a bit complicated, but we kept as much of Allison words and phrases as possible because of the insights they can provide.

Model I: Rational policy

For some purposes, governmental agency behavior can be usefully summarized as action chosen by a unitary, rational decision maker: centrally controlled, completely informed, and adhering to the agency culture, values, legal and regulatory guidance in a way that is expected of the agency. But this rational model, described more fully before, is a simplification that must not be allowed to conceal that most government agencies consist of a conglomerate of semifeudal, loosely allied organizations, each with a substantial life of its own.

Analysts attempt to understand happenings as the more or less purposive acts of unified agencies, with all of the staff having similar motives and methods of acting and reacting. For these analysts, the point of an explanation is to show how the agency could have chosen the action in question as a rational attempt to solve a problem or respond to a situation. Explanations produced by particular analysts display very regular, predictable features. This predictability suggests a subculture with set ways of facing and reacting that is both evident when looking within the agency as well as watching it from the outside. It should be no surprise that the rationale model is not realistic; it is just a way of thinking that helps frame decision making. These regularities reflect an analyst's assumptions about the character of the puzzle, the categories in which problems should be considered, the types of evidence that are relevant, and the determinants of the occurrences and the clusters of related assumptions. However too often we hear individuals say "FEMA or Department of Homeland Security of any other agency" will act in a certain manner, as we might expect if logic and accurate and extensive information were the only criteria involved. However, sometimes in certain uncomplicated situations, say something like "the Secret Service will never allow the President (any President) to venture outside the White House without complete security coverage" the use of Model 1 can be a quick and accurate predictor of Secret Service action.

Model II: Organizational process

If Model I is more of a framework for considering agency decision making, under rare, but unusually rational circumstances. Model II is a more realistic description of the background of much agency decision making. Agencies perceive problems through organizational sensors. They define alternatives and estimate consequences as the agencies process the information they receive and expect to receive. Agency behavior can therefore be understood according to a second conceptual model, less as deliberate choice of leaders and more as outputs of large organizations functioning according to their various standard operating procedures. This assumes that for large agencies there are smaller suborganizations that have special responsibilities for some specific problem areas.

To perform complex routines, the behavior of large numbers of individuals must be well coordinated. Coordination requires the use of standard operating procedures (SOPSs). To be responsive to a broad spectrum of problems, governments consist of large agencies among which have primary responsibility for particular areas is divided. Basically, it is the SOP rules according to which things are done that keep the organization running well. Because they are performing complex routines, the behavior of large numbers of individuals must be coordinated. Coordination requires SOPS: Assured capability for reliable performance of actions that depends upon the effective behavior of hundreds of persons requires established "programs." But organizations do change, learning to address old problems more effectively or new problems that are outside of existing SOPs. Dramatic organizational change also occurs in response to major crises. So, SOPs change in response to learning as well as to crises. Of course, not all change is useful. FEMA was changed and minimized by the birth of the US Department of Homeland Security, but it actually benefited (in a longer term perspective) by its failures in responding to

FIG. 5   Each disaster offers opportunities to learn for decision makers. Problems occur when each disaster has new decision makers repeating old mistakes.

Hurricane Katrina with adequately trained staff and resources, and was significantly strengthened after Katrina, only to fail miserably in its response to Hurricane Maria in Puerto Rico in 2017. Not surprisingly, after the massive failures at Maria, FEMA was again the subject of renewed interest and saw its staffing enhanced considerably (Fig. 5).

The wise Emergency Manager considers the SOPs, the agency culture and relevant staff of his or her own agency as well as those public or private agencies that they are dealing with as a starting point, along with the available history of events similar to the ones being addressed. Of course, the Emergency Manager needs also to be sensitive to different agency organizational processes in play at the time, if there are colleagues able to quickly provide a historical perspective.

Model III: Bureaucratic politics

In addition to reviewing how a public or a public organization should ideally, rationally, make decisions, the emergency manager also must consider how the organization's culture, traditions, and standard operating procedures will affect its decision making. If the last two conceptual models have been complex, they are appropriate preparation for discussing this last Model III which is based on internal, bureaucratic politics. The leaders who sit on top of organizations are not a monolithic group. Rather, each is, in his or her own right, a player in a central, competitive game. The name of the game is bureaucratic politics: bargaining along regularized channels among players positioned hierarchically within the government. Government behavior can thus be understood according to a third conceptual model not as organizational outputs, but as outcomes of bargaining games. In contrast with model I, the bureaucratic politics model sees no unitary actor but rather many actors as players, who focus not on a single strategic issue but on many diverse problems as well, in terms of no consistent set of strategic objectives but rather according to various conceptions. And personal goals, making governmental decisions not by rational choice but by the pulling and hauling that is politics (Ref. 47, p. 262).

The emergency manager as bureaucrat

The middle of a disaster response does not appear to be the best place to being thinking about the role of political leadership as opposed to the role of professional bureaucrats in carrying out governmental functions, such as emergency management, and how complicated those decision making roles can become. James Q. Wilson,[60] in his *Bureaucracy* paints a somewhat negative, but no doubt realistic discussion of some of the restraints on any federal administrator:

> One of the central themes is that management of government agencies is powerfully constrained by limitations on the ability of managers to buy and sell products or hire and fire people on the basis of what best serves the efficient or productivity of the organization. Laws and regulations limit how people can be hired, greatly reduce the chances of firing anyone, and surround the buying and selling of buildings and equipment with countless rules about fairness and procedure.

Marissa Martino Golden[52] takes a somewhat different, more employee-oriented viewpoint:

> Imagine that you are a career civil servant, a GS-15, and that you work for the Environmental Protection Agency (EPA). You have spent your career at the EPA enforcing laws designed to reduce pollution. Along comes a president (Note, this was written two decades ago) who is more concerned with reducing the regulatory burden on industry than with pollution, who confuses nitrogen oxide (a harmless substance) with nitrogen dioxide (a pollutant), and who states that Mount Saint Helens is a greater source of pollution than automobiles. Not only does he advocate policy change and appoint a new agency administrator, but in turn, that the new administrator shares his policy views and works with the Office of Management and Budget (OMB) to cut your agency's budget and curtail the issuance of new regulations. In short, all of a sudden, outsiders are not only telling you how to do your job but also trying to reorient your agency's mission.... Two questions arise: What would you do, and how?

Not surprisingly, frequently complex issues will arise during disasters precisely because a disaster is serious, because some people lose their homes, or even their lives, and because the disaster results in strong and expected emotions on all sides of almost all issues. Trying to complete emergency management missions in the face of all of this is not for the faint of heart. Especially in the light of a long-standing (though not universally accepted) political science tradition that has categorized government employees as just another group of self-seeking actors whose primary motivation is to seek higher budgets, larger missions, and staffs and the higher salaries that will likely follow—this is the budget-maximizing, self-serving bureaucrats described by Niskanen.[58,59] Unfortunately, this long-standing negative view of federal administrators is not a rare belief and can be quite strong.

Perhaps the most complicated issue an Emergency Manager must face is not strictly related to floods, hurricanes, the uses of Weapons of Mass Destruction, or other emergency or disaster situations. It is the fact that the higher up the chain an Emergency Manager serves, the more involved with political decision making he or she will become. "A first interpretation is that politicization is the result of prevailing balance between the political control that governments exercise over the administrative machinery and civil servants' involvement in the definition and implementation of public policy....In this sense, all civil servants are "political" because they are called upon to carry out political decisions, adapt them and explain them, in other words to accomplish work of a political nature that obviously is not limited to the mere application of legal or economic rules."[67] Moe[54] and Wood & Waterman[67] find that while political appointees can dictate the ideological coloring and overall direction of an agency, but these are often not central to the day-to-day issues they face, though large disasters are not "day to day" events, and will likely see higher levels of political involvement. In the complex, high stress environment of a disaster, all decisions, big or small, routine or not, are ultimately possible to become part of the political dialogue. Of course, many lower level decisions, for example, verifying that the specific objectives of the Incident Action Plan (IAP) regarding meeting scheduling and location are not necessarily the types of decisions in question, the objectives of the IAP itself could easily become "politicized" under some circumstances. And that is to be expected, though it happens only rarely.

As members of federal, state, or local agencies or consultants that assist those agencies, it is often the smaller, lower level organizational observations that will have outsized effects on decision making. Often experiential lessons can be learned just by asking "eye level" staff working on the issues. For example, though the Regional FEMA office staff and the Regional public health and medical response staff (US Department of Health & Human Services) work for different federal agencies, they often work with more often, meet with more often, and are personally closer to each other than to their counterparts in their Washington, DC, Headquarters offices. This is often a similar situation with other agencies that work with FEMA. Once a month, at the ten, Regional Interagency

Steering Committee sessions, the managers and key staff in each of the FEMA Emergency Support Functions (the 15 agencies that work with and support FEMA during disaster declarations made under the Stafford Act), meet, report on each other's activities and suggest ideas for more effective cooperation during disasters and emergency situations. These meetings are often followed by after work drinks. In this same vein, federal agency staff that work with state agencies in disaster preparation also tend to develop close professional and personal relationship that are not always understood or appreciated by the federal home offices. (With the exception of the "wise feds" who know of an appreciate the benefits of those relationships.)

It is a wise Emergency Manager who works hard to view the political sector as a group of powerful actors with complex needs and viewpoints, but one that has final authority for all major decisions before, during, and after major disasters. The Emergency Manager also needs to keep in mind that political actors as individuals or as members of groups have the same distorted (at least initially), bounded decision making capacities, and are subject to the same types of limited memory abilities, cognitive biases, and noise as are Emergency Managers acting alone or in groups. Following, we'll also cover a situation that at first seems "too far out to be real," but unfortunately is not, and that is the skill that must be honed to locate and then work with sociopathic managers. As we will discuss, the upper levels of corporate and even the military world are peppered with these unusually directed and certainly narcissistic individuals that can be dangerous to the mission as well as to your careers.

Although they are normally regarded as low-level public employees, the actions of most public service workers (especially Emergency Managers during disaster) constitute the services "delivered" by government. Moreover, when taken all together the individual decisions of these workers become, or add up to, agency policy, whether by FEMA, US Environmental Protection Agency, the National Disaster Medical System, or the US Army Corps., all key participants in disasters. During disasters, staff at federal (especially federal), state, or local agencies have considerable discretion regarding the amount, the location of deployment, and many other situations related to the goods and services provided. The choices of which agencies or organizations and individuals to work with is also a set of choice that can, in practical terms constitute policy choices. Over time, depending on where the Emergency Managers works, the outlines of his or her response virtually are always outlined in advance but an individual choice of a certain purveyor of food service, cleanup contracts, or communications professionals are choice that can influence the quality of the outcomes, from the view point of those affected, as well as the pricing of the goods and services.[56] And further, as Aaron Wildavsky and Jeffrey L. Pressman[57] remind us, the law as passed by Congress, the regulations implementing the law as developed by a particular political administration, and the "street implementation" by the on-the-ground administrators, leave quite a bit of room for the on-the-ground bureaucrat to administer as he or she sees fit. And, as long as private industry, local or state agency pushback is not experienced to any great measure, the on-the-ground decisions form the core of the program. These findings apply to FEMA as well as to each of the local and state agencies or other organizations that appear at disasters.

Can we improve our decision making?

It is disturbing, to say the least, to read Kahneman[29] stating that a considerable part of our thinking apparatus, the part that generates our immediate intentions, is not readily educable. But that is where his and Amos Tversky's considerable research has led them. In Finkelstein, Whitehead, and Campbell[225] considerable experience, we see the same findings:

> Unlike other writers (Often in the business/management area) in this increasingly popular field we believe it is impractical for us to correct our own mental processes. The brain's way of working makes this solution particularly difficult. Hence, when there are red flag conditions, we recommend safeguards that are external to the individual. We describe four safeguards; each helps to strengthen the decision process, so that the influence of distorted thinking is diluted or challenged (These are addressed somewhat more fully later.)

Except for some effects he attributes mostly to age, Kahneman finds his intuitive thinking is just as prone to overconfidence, extreme predictions, and the planning fallacy as it was before he made a study of these issues. Even when the cues to likely errors are readily available, errors can be prevented only by the enhanced monitoring and effortful activity of System 2.[27] As a way to live your life, continued vigilance is not necessarily good, and it is impossibly impractical; and of course, constantly questioning our own thinking would be impossibly tedious. The best we can hope for is a compromise: learn to recognize situation in which mistakes are likely and try harder to avoid significant mistakes when the stakes are high.[20] A major premise of his book was that it is "easier to recognize other people's mistakes in thinking than our own," an observation the authors turn to repeatedly in this book.

Kahneman came to his conclusions after lifelong study and research but he also observed the inherent weakness of decision making in his personal life. As a child he observed his father and many other educated Jews decide, with often tragic and deadly results, that the Germans were not going to repeat the types of mistakes that led to their decimation as a nation and often as individuals as well during World War II. Later in life, as a psychologist employed by the then-young but critically important Israeli Army, he ended up minimizing the ability of those selecting the Officer Corps from using personal judgment in the selection of officers, because the personal decisions of even those familiar with the military were almost always wrong. The reasons for these errors in judgment were eventually discovered (after much study and analysis), widely shared, and consistent. And often, even when biases or mistakes were pointed out, they still were repeated.

We will end this section on improving our decision making, difficult though that may be, by first by referencing T. Wilson[223]:

> The irony....is that they underestimate how valuable feelings are to thinking and decision making. It is now clear that feelings are functional, not excess baggage that impedes good decision making. Yes, there are times when emotions blind us to logic and lead to terrible decisions. In a fit of passion people sometimes do run off with the drug addled leader of a motorcycle gang. More commonly, though, our feelings are extremely useful indicators that help us make wise decisions. And a case could be made that the most important function of the adaptive unconscious is to generate these feelings.

T. Wilson[223] opens a section of his classic book on decision making with the title "Do Good, Be Good," and the he quotes Aristotle, "We acquire [virtues] by first having put them into action...we become just by the practice of just actions, self-controlled by exercising self-control, and courageous by performing acts of courage." In other words, certainly less eloquent and wise, if we want to become an effective decision maker who attempts to overcome his own emotional propensities, the effects of the moment, and any irrational predispositions that may affect us; we need to start trying to become that type of decision maker in our own lives.

Finally, we reference Kahneman and Tversky, regarding the difficulties of improving our decision making processes. Kahneman and Tversky were unwilling to extend their findings as sure-fire, "how to" rules for improving decision making. However, Michael Lewis[24] despite the systematic errors in decision making we are all subject to, they stress repeatedly that the rules of thumb that the mind uses to cope with uncertainty often worked well, but sometimes they didn't. And these specific failures were both interesting in and of them themselves and revealing about the mind's inner workings. Their work together (Tversky and Kahneman) stressed the usefulness of external mechanisms and procedures such as little mathematical formulas or summaries to help us avoid the systematic errors in decision making to which we all are subject, and which are so hard to avoid, even when we are keenly aware of them. Though, when Tversky and Kahneman worked as psychologists for the Israeli military they did frequently attempt to figure out rules, procedures, and even changes in selection criteria for placement of personnel that demonstrated that their own decision making processes, especially when full of introspections, can aid in the decision making processes of others.

Improving decision making by attempting to avoid some of our brain's limitations

Kahneman, in a somewhat more hopeful vein mentions times when awareness of your own biases can contribute to peace in marriage, specifically when each partner estimated what percentage of "home keeping tasks" each partner contributes. Though each partner tends to overestimate his/her percentages of task, still, the consideration of the types of tasks does improve at least this aspect of decision making.[20] These next paragraphs recount some additional ways in which some of the major researchers and authors have proposed to improve our own thinking. As you will see the only concept that appears in all of the various theories is the belief that it is really hard to improve our own thinking processes, but that there is much to be gained from developing systems and procedures to externally bolster decision making.

Using simple mathematical formulas to improve decision making

Based on earlier work on the value of simple formulas (algorithms) being more accurate predictors than professionals, Kahneman, Rosenfield, Gandhi, and Blaser[26] extended this same principle to professional decision making and found algorithms more generally effective than professional judgment. They found that professionals in many organizations that are assigned arbitrarily to certain kinds of cases, for example, physicians in emergency rooms, appraisers in credit ratings agencies, loan underwriters, and many others are not consistent in their decision making, making different decisions in similar circumstances, challenging the long-held values of "professional judgment." In the same vein Atul Gawande, who will be addressed in more detail below, has used check lists to assist and "routinize" many kinds of decision making that might otherwise be hampered by biases or noise. The problem as Kahneman, Rosenfeld, Ghandi and Blaser clearly describe is that humans are unreliable decision makers, too often strongly influenced by irrelevant factors such as by their current mood or whether or not they are hungry. This chance variance is called *noise*. Some jobs are noise free such as clerks at post offices or banks who perform complex jobs, but do so under strict rules that limit subjective judgment and guarantee, by design so that identical cases will be treated identically. Unfortunately, for even professionals whose judgment has been highly prized, studies have demonstrated a surprising lack of conformity, decision by decision, whether it be assessing biopsy results, valuing stocks, appraising real estate, sentencing criminals, or in many other situations.

Neuropsychologist Dean Bounamano[33] recognizes the same general types of brain flaws and "bugs" as he calls them, as many others. From the first person to tie a string around his finger to remind himself of something to the fact that our cars beep at us if we leave the lights on, we have developed a multitude of strategies to assist us in many facets of life. Buonamano favors variations of the same general types of "nudges" that economist Richard H. Thaler[226] has favored to push or lead us into sounder decision making. The unavoidable conclusions are that professionals often make decisions that deviate significantly from those of their peers, for their own prior decisions, and from rules that they themselves claim to follow. Since it has long been known that predictions and decisions generated by simple statistical formulas (algorithms) are noise free and are often more accurate than those made by even highly skilled professionals, Kahneman, Rosenfield, Gandhi, and Blaser recommend their use instead of professional judgment when possible, recognizing that algorithms will be operationally or politically infeasible in some instances. But, recognizing that substituting simple algorithms for individual or group decision making will not be an easy change to institute for most agencies or organizations. They recommend training; roundtables that review past decisions structured to prevent spurious agreements because participants quickly converge on the opinions stated first or most confidently; user friendly checklists; and carefully formulated questions to guide them, if time allows. Of course, the authors were addressing mainly institutional decision making at the corporate level, where there is time for training, roundtables, checklists, and question lists as aids in individual and group decision making (Fig. 6).

The Emergency Manager operates under tighter time lines than those faced by most corporate decision makers, but generally there is time to review predeveloped response checklists; small, well designed, decision roundtables; and predeveloped question lists (even the most experienced Emergency Managers can forget things in the earliest parts of a response). And there is no doubt that many logistics and financial issues can benefit from simple algorithms although some have already been developed, such as the knowledge of how much water or food is needed, on average, for each person affected in a flood or other event. The value of these types of preparations is twofold: First, they convince Emergency Managers to think more about the art of decision making itself, rather than just the substantive issues dealt with, and Second, they stress ways to minimize the weakness in decision making that become more evident the more the issue is studied. None of this negates the benefits of the knowledgeable

FIG. 6   Discussing what is going well and where we can improve is a constant cycle.

and experienced Emergency Manager using his or her informed intuition for some quick decision making, using the OODA Loop particularly
regarding using multiple information sources and constantly striving to
develop new or additional models of thinking, when that is called for. But
even then—watch out.

Applying some of the wisdom of Allan Jacob's to improve decision making

In a previous section we have presented some of the key findings of
Allan Jacobs in Improving the *art* of decision making (to use his term).
Jacobs stressed the strong tendency of individuals to join groups and
then to begin adopting the groups' perspectives, reality based or not, con-
sciously as well as unconsciously, and even its emotional response to certain types of situations. Based on one's ability to join or quit a group, one
does exert some control of how some of his or her emotional responses
are created and strengthened or weakened. Joining formal and informal
groups of truthful and right-thinking people is one of the best ways in
which we have to steer the development of our own minds as developing
the habit of trying to be skeptical of our own reasoning processes. Jacob's
also resurrects the familiar warning against letting any strong and existing emotions dictate our thinking and our decisions, he advocates waiting
for a few minutes before making any critical decision. And always, to the
extent practical and possible, we can always attempt to be aware of our
decision making processes and any restraints on them.

Applying Finkelstein, Whitehead, and Campbell (FWC) methods to improve decision making

FWC have studied some of the biggest national blunders such as the Iraq War to analyze the failure of leaders to make good decisions. The deeper they went into decision making in new or unfamiliar circumstances (even in well-known areas) they found "flawed decisions abound." Two factors were in play: an individual or group error of judgment, and a decision process failing to correct it. They found that complex decisions, involving interpretations and judgment are difficult to get right, under any circumstances. They found four conditions when flawed thinking is likely, where even the experienced can get it wrong, calling these *red flag conditions*. FWC sought external *safeguards* balancing flawed emotional tags underlying these red flags.

The four red flags:

First are *Misleading judgments*. Faced with unfamiliar inputs, especially if they appear familiar, we can think we recognize something when we do not. Our brains contain memories of past experiences, each tagged with certain emotions, connecting with inputs we are receiving. Think about vertigo or fear of dogs; we may know rationally there is no danger, but still may be unable to control our fears. Past experiences may not be a good match with the current situation and thereby mislead us. Another example is when thinking has been primed before we receive inputs, by, for example, previous judgment or decisions connecting with the current situation. If these are not applicable to current situations, they disrupt our pattern recognition, causing us to misjudge information received.

Second, we call these *misleading prejudgments*. Emotions attached to misleading experiences or prejudgments can supply more emphasis in our thinking than appropriate.

Third are *inappropriate attachments* to failed ideas, for no reason other than we enjoy them. As an example, FWC cites Samsung CEO, Lee Kun Hee's disastrous attempt to build an automobile with no experience. He loved automobiles, rejecting rational reasons why Samsung should not make them, his emotions producing strong tags leading to an irrational, costly mistake. Inappropriate attachments easily arise from family, friends, favored communities, and beloved objects.

Four, *inappropriate self-interest*, are actions that unfairly benefit us. FWC cite a study of medical residents finding 62% of respondents said "promotions don't influence my practice" meaning when prescribing they were not influenced by promotions of drug makers. In fact, only 16% of them believed the statement true of other physicians. As it turns out, studies confirm promotions do affect prescribing. These errors would not be a problem if we made decisions in controlled ways with checks and balances. But we do not.

To guard against the risk of errors in thinking, FWC developed four safeguards that are external to decision making. The four safeguards are as follows:

One, development of Experience and Data in important areas.
Two, Group Debate and Challenge session instituted in important decision making areas.
Three, a Governance Team as a backstop to flawed ideas that have made it through to final decision levels.
Four, Monitoring the decision process at all levels. These procedures have not been detailed as they are mostly not appropriate to the agencies involved with Emergency Management, but the errors of decision making so aptly described by FWC are, so they were covered more deeply.

The value of checklists in decision making

Atul Gawande studied complex decision making situations in which a checklist would have been very helpful, if not indispensable to good decision making. A major example was the aftermath of the failed Hurricane Katrina response and recovery missions which he correctly ascribed to a lack of local, federal, and state competence and coordination. Gawande was unaware of an even greater restraint on good decision making. The ranks of the Senior Executive Service and G.S. 15 managers at FEMA's Headquarters office were purposely thinned out by a reduction in force, unfilled senior level vacancies, staff transfers and resignations that occurred under the first FEMA Administrator appointed by President George W. Bush, in the mistaken assumption that the ensuing staff shortages would be effective cost containment measures. Under other circumstances, a robust FEMA Directorate staff could have quickly located and predeployed buses, helicopters, bottled water, medical supplies, food, transportation/rescue assets, and so on, while beginning communications with the devastated local and state emergency management leadership, under existing "life safety" authorities and practices. As experienced professionals they would have begun predeploying assets before the storm's landfall (though some federal ESFs actually did this, as previously discussed). Nonetheless, response checklists of the kind advocated by Gawande, if they existed at local, state, and federal levels would have facilitated a much better response and recovery mission than actually occurred.

Gawande used the successful examples of National Transportation Safety Board-mandated airline flight safety checklists, addressing the complexity of developing huge sky skyscrapers, the growing numbers of surgeons using checklists during surgeries, and many other examples to demonstrate that the growing volume and complexity of so many technological areas will benefit from the development and uses of checklists.

His basic theory is directly applicable to Emergency Management. While it is true that an experienced and skilled Emergency Manager when dealing with a flood, for example, quickly would begin to address the need for massive amounts of bottled water (but not stored in warm warehouses!) to be dispatched to areas near the flood zone, the need for water testing by local health departments, or by the Centers for Disease Control on site, the need for mobile, generator-powered renal dialysis (since electric power would likely be out) stations to address patients needing renal dialysis after their hospitals have been taken off line, the need for shelters and medical personnel to staff them, cell tower repair and mobile repeaters to begin to reestablish cell phone communications, and so many other items. However, memories are fallible, staffs change along with conditions, to say nothing of flooded roads, brittle diabetics needing insulin right now, and other issues needing immediate attention. Even relatively rote-list oriented items addressing flood-related needs would benefit by existence of relatively short, yet detailed, prioritized response and recovery checklists at local, state, and federal levels.

Gawande observes that there are a number of key decisions when making a checklist, starting with the need to define exactly when the checklist is to be used (though sometimes that is very obvious). First you must decide whether you want a DO-CONFIRM checklist or a READ-DO checklist. With a DO-CONFIRM checklist, team members perform their jobs from memory and experience, often separately. But then they have to stop. They pause to run the checklist and confirm that everything that was supposed to be done was done. With a READ-DO checklist, on the other hand, people perform tasks as they check them off—it's more like a recipe. So for any new checklist created from scratch, you have to pick the type that makes the most sense for the particular situation. And you have to keep the checklists as short as possible by focusing on the killer items, "the steps that are most dangerous to skip and sometimes are overlooked nonetheless." Gwande's book, *The Checklist Manifesto: How to Get Things Right* is one of the books that the authors have recommended as having a place on the book shelf of a wise Emergency Manager.

As a last thought, the Emergency Manager can benefit from an awareness of the key thoughts presented in the many books and pamphlets on Mindfulness. Though these often covers such issues as meditation, being attuned to your body and similar "semi-new age" thoughts, which can be helpful in themselves, but may not be on the main agenda of most Emergency Managers, there is something of real value for people's thoughts as Emergency Managers. Each definition of Mindfulness shares at least one concept, an extended awareness that is now referred to as mindfulness. It means we should be aware of our bodies, that we constantly be mindful of what we are doing, our surroundings, mental and physical, and of how we should best react to it all. This is how the

Emergency Manager should operate, having developed the habit of always being aware but skeptical of his or her decision making process, being aware of the thoughts and circumstances that might be affecting the decision making process, and that a similar process is going on with each individual and group we come across as a disaster response and recovery develops. The decision making process can be helped with checklists, with the acquired habits of looking for red flags in decision making and a realistic assessment of the value and the limitations of groups in decision making.

Evolutionary approaches and game theory

After Sociobiologist E. O. Wilson[60] studied Darwin's theory of "the survival of the fittest," that over time, the species that were best adapted to their environments survived while those that did not adapt as well perished, he began applying Darwin not only to physical traits, but to human social and cultural traits as well. With this "applied" theory of Darwin, E. O. Wilson began to show that the most well adapted social and cultural patterns persisted and changed as the human environment changed. Of course seeing the long neck of a giraffe as an effective adaptation making it easier to eat leaves from the top of trees, where few other animals could reach, is a much more evident application than viewing the adaption of human social and cultural traits. But as Wilson and other sociobiologists such as Wilson D. Hamilton[61] demonstrated, many human social and cultural traits have evolved over time. As we will see below, there is a place for some knowledge of evolutionary biology in emergency management.

As evolutionary approaches multiplied and were used to explain a growing number of human behavior patterns, Col. Boyd began recognizing sociobiological reasons for some types of persistent and widely shared conduct, though we could find no references to his studying game theory. We have little doubt if he had lived longer, Col. Boyd would have "adapted" the evolving game theories to these sociobiological approaches to the many models he had already absorbed. This brief discussion of evolution and sociobiology leads into Robert Axelrod's use of these theories in his approaches to develop more effective models of cooperative decision making.

Over 40 years ago Robert Axelrod began using computer game simulations to determine the most efficient way for individuals to "win" in head-to-head competitions or negotiations of many kinds. He had long felt that reciprocal cooperation among individuals (dealing with each other fairly and kindly) would result in the highest potential outcomes for both. He wanted to test his "gut feeling" and chose the classic Prisoner's Dilemma to do that.[62] The Prisoner's Dilemma game allows the players to achieve mutual gains from cooperation, but it also allows for the possibility that one player will exploit the other, or the possibility that neither will co-operate. As in most realistic situations, both players do not have strictly opposing interests. When Axelrod began his studies those economists and political scientists who felt individual decision makers acted rationally in pursuit of their individual goals, also felt that most individuals in situations such as the Prisoner's Dilemma could or would not cooperate. In other words, the first player would not act cooperatively on his or her first turn, and neither would the second player on his turn, and on and on.

Axelrod's project began with a simple question: When should a person cooperate, and when should a person be selfish in an ongoing interaction with another? Should a friend keep providing favors to another friend who never reciprocates? Should business associates provide prompt service to another business associate that usually pays late, or is about to be bank-rupt? In a wider perspective an Emergency Manager is always making decisions in groups and sometimes in one-to-one negotiations with indi-viduals that are not known and familiar, but who still must be negotiated with successfully. So, even if decision making tools can be sharpened, or at least more effectively observed in others, the format of a negotiation about decision making choices is familiar to us all. Axelrod received sixty-two entries from computer hobbyists, evolutionary biologists, political scien-tists, economists, sociologists, and others. To his "considerable surprise" the winner was the simplest strategy of all the programs submitted. TIT FOR TAT, TIT FOR TAT. It was merely the strategy of starting by being co-operative, and thereafter doing what the other player did on the previous move. On your first move you cooperate, if the other player cooperates, the string will tend to persist. If the second player does not reciprocate with co-operation, the first players reciprocate with a negative or selfish response.

Cooperation in biological systems can occur even when the participants are not related or are unable to appreciate the results of their own behav-ior. What makes this possible are the evolutionary mechanisms of genetics and survival of the fittest. An individual able to achieve a beneficial re-sponse from another is more likely to have offspring that survive and that continue the pattern of behavior which elicited beneficial response from others. Thus, under suitable conditions, cooperation based upon reciproc-ity proves stable in the biological world; potential applications are spelled out for specifics aspects of territoriality, mating, and disease. The conclu-sion is that Darwin's emphasis on individual advantage can, in fact, ac-count for the presence of cooperation between individuals of the same or even different species. As long as the proper conditions are present, coop-eration can get started, thrive, and prove stable.[62] The analysis of the data from the tournaments demonstrates four properties that made decision rules used in the tournament successful. Axelrod summarizes them,[62] "in a nutshell": The evolution of cooperation requires that individuals have a sufficiently large chance to meet again so that they have a stake in their future interactions. If this is true, cooperation can evolve in three stages.

(1) The beginning of the story is that the story of cooperation can get started even in a world of unconditional defectors;

(2) The middle of the story is that a strategy based on reciprocity can shrive in a world where many differences kinds of strategies are being tried;

(3) The end of the story is that cooperation, once established on the basis of reciprocity, can protect itself from invasion by less cooperative strategies.[62]

All throughout the tournament those using the successful TIT FOR TAT, TIT FOR TAT strategy made a point of engaging with their opponents in a progressively positive and helpful manner, as often as they could within the structural and the time constraints to the game. This idea of building confidence by continually progressive "constructive engagement" is the same strategy that has successfully been used in diplomatic engagement for hundreds of years. Usually it works.

Based upon Axelrod's' tournament results, four simple suggestions summarize what he learned from the tournament players: do not be envious of other players' successes; do not be the first to defect into selfishness; reciprocate both cooperation and defections; and do not be too clever. The existence of all that has been learned should serve as a warning that the facile belief that an eye for an eye is the best strategy. And above all, each player takes into account what he or she thinks the other player will do, so it is always wise to speak and act in a manner that others can easily interpret as cooperative, but, as noted, after cooperation is extended, if it is refused, the initial co-operators' retaliation can have the positive effect of drawing a cooperative effect the next time. Even a quick reading of the material just presented shows that the Emergency Manager is best off being known as a cooperative individual, but not one who is always a "push over." Of course, most Emergency Managers have already absorbed these lessons without studying game theory, but game theory does affirm what common sense and good emotional intelligence tells us, it is better to be known as cooperative than not.

At this point the reader should be getting the message that the life of an Emergency Manager is anything but simple and straightforward. Mitigation, preparedness, response, and recovery are easy to say but consider the players involved in each. For each category there are multitudes of players in the public and private sectors. If the Emergency Manager controlled everything necessary to mitigate, prepare, respond, and recover from a known hazard, life would be simple. The Emergency Manager must not only make critical decisions that can impact life safety of hundreds if not thousands and yet they also must make those decisions in a manner acceptable to all players in order to accomplish the mission. Not an easy ask for anyone.

Decision making in emergencies, disasters, and catastrophic events

Chapters 1 and 2 introduced the science as well as the practices of the art of decision making as studied by the military, behavioral economists, political scientists, business and management experts, and most of all by psychologists, blending these fields together for a more complete understanding of decision making. Chapter 3 combines the brief introductory chapters with a consideration of decision making through history and leads directly to this chapter's extended discussion of decision making in specific, applied emergency and disaster situations. Chapter 4 references the disaster literature, classic texts, after-action reports, and a wide variety of mass and social media sources. It also includes ideas culled from the experience of the two authors in over 80 years of cumulative emergency management experience in order to derive perspectives in emergency management decision making that might not be presented elsewhere. Chapter 4 is pragmatically oriented. It places local, state, and federal emergency management decision making in the traditional four phases of a disaster: preparedness, response, recovery, and mitigation in the types of events that are rarely, as well as routinely encountered (e.g., weather events, national special security events, etc.[63]). Traditional emergency management definitions will be used whenever practical as opposed to the many revised and changing regulatory and statutory provisions. Of course, the definition of an "emergency" if not of a "disaster" is, to a great degree, determined by the type and intensity of the media coverage surrounding the event and of how the event strikes the viewers. As Elisa Gilbert, writing in the *New York Times Magazine* wisely observes, "In other words, our collective reckoning of what is or isn't a disaster entails publicity. Disasters are news because they are news.… They are public events that feel personal. Its why people remember where they were on 9/11, and some even claim to remember watching the first plane strike on TV that morning—an impossible memory, since only the known footage of the event was captured accidentally and was not on the news that day."

Since so much of emergency management decision making takes place within the various emergency operating centers (and the various other names these decision making offices may be called at local, state, or federal levels), we will start with a basic description of the Emergency Operations Plan (EOP) which is the major product of these Emergency Operation Centers. The EOPs that are made are among the first public documents both defining how society considers the emergency or disasters, and also tend to mark society's initial response to that event. The Emergency Operations Plan will normally address 10 key areas (Fig. 1):

1. Local and State Descriptive Information
2. Locally Based Hazard and Risk Reduction
3. Communications
4. Disaster Annexes

FIG. 1   Policy makers and politics add complexity to decision making for the emergency manager.

5. Staffing
6. Exercising, Review, and Evaluation
7. Safety and Security

Decision making in lesser emergencies, ramping up to disasters, and catastrophic incidents will be addressed with specific reference to floods, wildfires, tornadoes, hurricanes, winter storms, earthquakes, tsunamis, heat waves, technological disasters, and terrorist events. National Special Security Event Pre-Deployments (e.g., Meetings of the G-7, Republican and National Conventions, some huge sporting events such as Baseball's World Series or Football's Super Bowl, etc.) are relatively frequent, though potentially hazardous events which can sometimes include preparedness and response actions in many of these areas. But wherever Emergency Managers find themselves during significant disaster events their initial decision making process is pretty uniform. Haddow et al.[2] find that almost all preparedness planning activities begin with a basic Hazards Risk Management (HRM) process used to systematically identify and assess risk. As a lead in to Chapter 3, we presented a fuller decision making matrix based on military sources, which have through the years studied "on the ground" decision making more intensely than any other organization. If an Emergency Manager asks a military decision maker why they are so confident in their processes, and have taken it so seriously, the standard answer is usually something like this: "Because lives depend on good decision making in the military in a way that they do in no other place." Regardless of the agency or institution almost all HRM methodologies include the following steps:

- Identify the hazards
- Assess the risks for each hazard identified
- Analyze the hazards risks in relation to one another
- Treat the hazards risks according to prioritization

In order to best analyze and assess a hazard, a concise hazard profile needs to be developed to place the hazard in a local context with appropriate supporting information. The following are examples of information that are often investigated and that, to the extent that they have been developed by the Emergency Manager prior to the events in question, will facilitate the quick and comprehensive decision making that often must accompany response to serious emergencies, disasters, and catastrophic events.[2]

- General orientation overview of the hazard
- The location of the hazard within and surrounding the areas of study and the spatial extent of its effects
- The duration of an event caused by the hazard
- Seasonal or other time-based patterns followed by the hazard

- Speed of onset of an actual hazard event
- Availability of warning to the hazard

Of course, this is all subject to the oft-repeated FEMA "law" that once a disaster occurs, the plan has lost most of its value because what has been planned for will rarely occur in any form near to that which has been planned for. Identifying the hazards is achieved by using a variety of methods, including historical study, brainstorming, scientific analysis, and subject matter expertise (SME). Fortunately, newer technologies are available when "mining" data to include Automated Data and newer Artificial Intelligence processes. All can be helpful, but during an actual disaster these newer processes can tend to overburden staff as upper management, particularly politically appointed upper managers, thirst for data, and information to report up their chains. Preexisting protective "rules" can be helpful, but top managers approve rules and can be driven so hard from their own chains that data must be provided, regardless of the practical efficacy of doing so. Meeting these needs while responding to an event points to a constant theme of emergency management—locating and using the best of the newer technologies while retaining and even enhancing some older, proven processes. Sometimes the tradeoffs are not as simple as they may initially appear. For example, a decade and one half ago, when FEMA adopted use of Blackberries for phone, texting and email, Emergency Managers in the field could, quite literally, "Run the disaster out of their Blackberries." As time passed most federal agencies began supplanting the trusty Blackberry with more powerful, multiuse IPhones and other brands, despite the fact that none had the quick key board capacities and stable email capacities of the earlier Blackberries. Unfortunately, Blackberries were never upgraded in a timely manner, so even the newer, upgraded, models are rarely used as disaster response communications equipment (though some federal staff still retain them because of the superior keyboards).

This text will emphasize the response aspects of emergency management decision making, as this is the time, especially true for health and medical, during the "golden 24 hours" when most disaster casualties are most effectively addressed, and need to be located, treated, transported, or evacuated. The biggest successes and the biggest failures of judgment usually occur in the golden 24 hour period when an Emergency Manager's individual strengths and weaknesses in decision making, often under the pressures of time stress and emotions are made. Of course, the same individuals who make response decisions, are not necessarily the same individuals making decisions in preparedness, mitigation, and recovery situations, though FEMA, to a great extent, has moved much of its previous mitigation concerns (for now at least) to agencies and entities somewhat outside the direct emergency management community.

Emerging trends and history affecting emergency management decision making

Background

Natural disasters are becoming more frequent, more severe, and affect more people than ever before; reasons vary but include climate change, population growth, and shifting habitation patterns that include living along ocean, river, and lake shores, "building and rebuilding" in risky areas even as we avoid the policies and investments that would help mitigate the threats".[228] As a result, developing the tools, processes, and best practices to manage natural disasters, technological events, and terrorist events more effectively is becoming an increasingly important global priority. Terrorism response, primarily as an enforcement activity, is not an emphasis of this text. We emphasize events that accompany or result from terrorist events such as evacuation, safety, medical treatment, environmental and public health, infrastructure, and related emergency management concerns. Though the term "All Hazards Approach" is constantly being changed, emphasized, and deemphasized by the various Presidential Administrations, the tools available to the Emergency Managers, as vast as they are, remain relatively fixed (this has normally been referred to as the All-Hazards Approach, where the Emergency Manger his his/her set of tools which is adapted to virtually all disasters or terrorist situations). However, despite record advances in using enhanced information and communications technology for disaster response and humanitarian assistance, many desirable, coordinated, international solutions simply do not exist yet[66] so the focus is more often nation-based activities, though there is a growing international mutual aid and SME potential under a growing variety of information and asset sources.

Significant emerging trends continue to shape how emergency management is funded, practiced, and organized at the US federal, state, and local levels. Changing patterns of emergency management decision making are reflected in these changes. Understanding the pulls and pushes affecting emergency management decision making is important because at different times the concepts of emergency management have been and will be, of necessity, applied differently. The definition of *emergency management* itself can be extremely broad and all-encompassing; but unlike other, more structured disciplines, for example, public health, medical systems, or environmental management, emergency management has expanded and contracted in response to naturally occurring (e.g., Hurricane Katrina in 2005) or significant induced events (The World Trade Center bombing in 1993 and in 2001), congressional desires, and leadership.[1] Emergency management is not based on gradually developing pools of scientific knowledge in the field (though this does exist to some degree).

Emergency management remains more fluid, depending as much on changing networks and funding sources as on its foundations principles, such as resiliency initiatives, the location of most mitigation funding and policy efforts, all hazards planning and response. And, as Waugh and Tierney[2] remind us:

> Some of the problems that Emergency Managers have experienced with homeland security and policies and programs reflect basic differences of opinion about the role and function of emergency management, and about public participation in decision making, transparency in decision making, and information sharing with other agencies and the public. Just as federal priorities often conflict with local, state and regional needs, the federal approach to emergency management often conflicts with the approach of professional Emergency Managers. In particular, the centralization of decision making encouraged by federal authorities conflicts with the decentralization of decision processes at the local level.[68,69]

Emergency management is as much a reflection of its times as it can be important in responding to the major events in a time period, for better or worse. There are a number of circumstances affecting Emergency Managers across the world, and certainly in the US, in the second decade of the century, these include:

(1) The 911 Attack and the Development of the US Department of Homeland Security,
(2) The Lessons from Katrina and the 2005 Hurricanes,
(3) The Extended, Nonfederal Government,
(4) Increasing Vulnerability to Global Disasters,
(5) The Professionalization of Emergency Management
(6) The Disadvantaged and the Vulnerable as a Special Concern in Emergency Management. Decision making perspectives will be woven into these discussions.

The 911 attack and the development of the US Department of Homeland Security

As a backdrop to decision making in various emergency, disaster, and catastrophic events we must recall that the momentous events of 911 (and the earlier 1993 attack on the World Trade Center that was a lead up to 911) reshaped emergency management in ways that are still having effects. Looked at in a wider perspective, the overall state/federal response to the event was adequate. It was the lack of local, interoperable communications systems that resulted in the deaths of many first responders that probably needn't have died. The few casualties that were alive were rapidly addressed with standard EMS procedures. The ongoing mission was mostly a disaster mortuary mission that focused on securing, identifying, and respectfully disposing of human remains, often in small and burned particles. In fact, this was the first, massive use of DNA identification procedures and marked the rapid transition of mass death events from the FBI to the Disaster Mortuary System portion of the National Disaster Medical System. For local, state, and federal emergency management decision makers, the response tilted toward the mortuary mission and the provision of immediate health care to injured first responders. But the 911 attack, as well as the earlier 1993 World Trade Center terrorist attack convinced Congress to shift the emphasis of emergency management from the more traditional, naturally occurring hazards to terrorism.

There were also at least two intel failures that may have limited the 911 attack: One of these had to do with a Minnesota FBI field request to search the possessions of Zacarias Moussaoui, a suspicious flight student and a colleague of Islamic fundamentalists, which contained information related to the 911 attack and information related to the preplanning for the attack. Apparently the Justice Department initially held the Minnesota FBI agents from opening searching his possessions. Another agent wanted to move his squad into an effort to find Khalid Alminhdar, an al-Qaeda operative loose in the country. Alminhdar flew the plane, American Airlines Flight 77, which crashed into the Pentagon.[70] Some agents later resigned after the OGC action, though one of them remained an agent who eventually told her story on the CBS 60 Minutes news show. The FBI agents were skilled and dedicated. Sadly, the event demonstrated that sometimes lawyers, feeling they are being "good" lawyers, missed the fact that the legal point did not outweigh the potential value of any Al Qaeda intel. Shouldn't the lawyers have sought to find a legal way to open investigations immediately, instead of interposing their view of the law over the value of the intel. One of the authors was acquainted with some of the FBI agents involved and appreciates the anguish they felt. The policies that were invoked in the earlier issues were initially developed by the Clinton Department of Justice to guide terrorist investigations; ironically, the Clinton Administration that left actionable intelligence for the incoming Bush Administration regarding a potential Al Qaeda air attack. The information made its way into the mass media, but quickly melted away. The information was neither taken seriously nor acted upon and some of it still remains classified.[71] The point of this is not to berate the Clinton or the Bush Administrations with the tool of perfect hindsight, but to stress a constant theme that professionals acting in their professional capacity are not to be fully trusted to make accurate decisions that relate to events outside of their professional fields of expertise. Unfortunately, with lawyers, doctors, and some other professionals, the "Hammer and nail" principle is always something to contend with. And, in the case of the powerful, new appointees of the Bush Administration, the overconfidence that can accompany those early and powerful political appointments is understandable, but nonetheless a constant threat to good decision making. But in absolute fairness, as the *National Commission on Terrorist Attacks Upon the United States* points out after a terrorist event what appears clear in hindsight, is always immersed in the issues of the day and will never appear as "clear' then, as it appears now.[72]

None of the changes that will be discussed had the long-lasting effects on decision making as the September 11, 2001 attack resulting in the development of the federal Department of Homeland Security (DHS). After the 911 attacks the world of emergency management became much more complex. The roles and functions of Emergency Managers did not change dramatically; terrorism was one focus as were naturally occurring and technological hazards, but now the attention of the public and its elected officials were focused on the newer, war on terror, while the "older" hazards of floods, fires, tornadoes, hurricanes, and earthquakes were still very much on the agenda, and if anything were intensifying to the extent that global climate change and other causes are involved.[67] The new and powerful DHS included not only FEMA, but the FBI, the CIA, the Secret Service, Immigration and Customs Enforcement, the Coast Guard and other, response type agencies. At the time of its development, all or part of 22 federal agencies were initially housed in the first Director, Tom Ridge's, Department of Homeland Security (DHS). Wisely the FBI and the CIA were quickly pulled from the huge agency. As an aside, in one of the federal regional offices that both authors were associated with, the first meeting of the Regional Advisory Council included, for the first time, representatives of all of the 22 federal agencies now housed in the new DHS. When those in attendance looked around the room and saw such a seemingly unrelated groups of agency staff, those in attendance burst out in spontaneous laughter. The Regional Director, not surprisingly, was miffed, at first not seeing the incongruity of it all.

FEMA remains submerged in DHS, with its primarily weather-related disaster focus being de-emphasized and minimized as DHS placed its funding and organizational emphasis on preventing and responding to terrorist events. FEMA budgets have never since risen enough to enable state emergency management agencies to pass through sufficient local emergency management funding so that locals can do their primarily weather and natural event preparedness, response, recovery, and mitigation jobs along with their terrorism missions. However, in fairness, many rural Emergency Managers have rightly scaled the terrorist mission, unless they are fortunate enough (or unfortunate enough) to have power generation infrastructure or other high-damage potential, though minimal risk, targets in their regions. The great progress that FEMA made with President Clinton's appointment of James Lee Witt, and the widely experienced "Southern Emergency Managers" he brought with him was to a great extent lost after the birth of DHS, especially after the shift of emphasis from existing natural and technological events to terrorism. Although FEMA was "re-supported" after the public embarrassment of Katrina and the the exception of the massive FEMA failures in Puerto Rico during Hurricane Maria in 2017, FEMA has performed relatively well in hurricanes and winter storms (Hurricanes Sandy). Its failures in Hurricane Maria were part judgment based, and part because funding levels did not allow for it to address more than one major event at the same time.

The Emergency Manager, especially at local, rural level, has only to meet flow-through FEMA-state and DHS requirements regarding antiterrorism, expecting to have few or no decisions to make regarding terrorism, save possibly having to help in a larger jurisdiction's preparations, or a state or FEMA training or exercise, at least, to date. Of course addressing the news and longer term issues of a terrorist event will still fall directly in FEMA's lap.

Hurricanes Katrina and Rita in 2004, 2005

The busy 2004 hurricane season and the embarrassingly poor responses to Hurricanes Katrina and Rita forced yet another reevaluation of FEMA priorities, funding, and organization; the major lesson of the 2005 hurricanes being that the nation's emergency management system was, once again, broken and that officials, along with Congress, needed to rebuild local, state, and regional capacities and resource levels to reduce hazards and to respond to disasters. The Hurricanes also vividly illustrated the political costs of failing to prepare for disasters as well as the fact that nearly ¾ of the deaths in most hurricanes are from storm surge and flooding, and not from wind damage. The severity of the Hurricane Katrina disaster cannot be minimized with nearly 2000 lives lost, mostly by drowning, in Louisiana, Mississippi, Alabama, Florida, and Georgia. Entire communities were wiped off the coast, and hundreds of thousands were displaced. Four levees breached in New Orleans in the storm, culminating in floodwaters of 15–20 ft covering 80% of the City. In an aside, this was the first major disaster that FEMA and other disaster employees used Google Earth to get sky view of disaster damage; and for the first time saw the small circles almost like donuts (composed of rubble where houses once stood in tact) that were sprinkled across areas that experienced Category 4 or higher hurricane winds. These images persist for the agency responders, including the authors.

In many respects, FEMA's failures were to be expected where it was too understaffed and "out of the loop" to absorb good fly-over information and begin to respond immediately. It was not just that the two previous FEMA Directors, Joseph Albaugh and Michael Brown were not fully up to their missions, other Directorate level staffs weren't either. FEMA's Logistics Chief and his staff were also newly installed in their positions and did not yet have good information regarding their own assets and response regiments and did not locate and quickly deploy buses and other key assets in a timely manner. In fairness, Brown also inherited staffing shortfalls in addition to his inexperience and lack of disaster credentials. FEMA, in 2005 during Katrina had only 2100 full-time employees and had its senior ranks further decimated by George W. Bush's first FEMA Director, Joseph Albaugh who did not replenish the Directorate leadership (Senior Executive Service, G. S. 15s, etc.) when he assumed his position. In fact, newly appointed FEMA Director Albaugh initiated a Reduction in Force (RIF) in the ten FEMA Regional Offices (that housed most of FEMA's direct response staffing experts) presumably as a misdirected "cost saving measure" during his tenure. FEMA Regional Office Directors wisely, but quietly, "pushed back" by not directly laying off staff, and instead by reducing unfilled FTEs and by other methods for minimizing the effects of this damaging initiative. Staffing has since grown, though not necessarily in numbers adequate to the tasks now facing the agency.

Because state and local officials have had to rely to a greater extent on their own often minimally resourced preparedness, response, recovery, and mitigation resources, mutual aid under the Emergency Management Assistance Compact (EMAC) was key to supporting Katrina's aftermath. The mission of MABAS (The Mutual Aid Box Alarm System), using the example of Michigan in this instance, was to coordinate the effective and efficient intrastate and interstate mobilization and deployment of fire, emergency medical services, and special operations mutual aid resources, during natural and man-made emergencies and disasters. The largest mutual aid network MABAS was extremely active in the Katrina mitigation efforts. MABAS started in Illinois in the 1960s and has spread throughout the Midwest and other states.[73,85]

It would be a mistake to categorize the Katrina federal response as a total failure. For example, the National Disaster Medical System under Director Jack Beal had seen longer term forecasts and deployed Disaster Medical Assistance Teams (generally 35–50-person medical response teams composed of at least one physician, paramedics, EMTs, nurses, communications, and logistics staff). These medical assets were deployed either to hardened sites in which to "ride out" the storm (hotels or high schools which are often built to withstand wind of 150 mph or more), or to sites away from the expected hurricane path. NDMS Emergency Coordinators were deployed with DMAT elements, and at FEMA and at various state and local operation canters, to develop situational

FIG. 2   DMATs deploying ahead of a potential disaster.

awareness, to conduct medical needs assessments, and to secure additional public health and medical assets as needed. The complex logistics involved in deploying (to say nothing of predeploying), NDMS DMAT assets cannot be minimized (Fig. 2).

A typical 35–50-person DMAT element can require approximately one half of a mile or more caravan of vehicles that will include staff, 53-ft trucks carrying generators, treatment tents, masses of pharmaceuticals, communications equipment, and more. It should also be noted that NDMS is not authorized to carry arms, so must depend on local, state, and federal sources for security, never an easy thing to do during a disaster. NDMS staff, unloading boxes of controlled substances in the middle of the night in New York City during Hurricane Sandy, found out to their dismay that they were located across a lot from an all-night Methadone treatment facility (with good luck, there was no tragedy).

Nurses, other HHS personnel, and personnel from 3 NDMS DMATs were assigned to the New Orleans Airport where they were responsible for perhaps the most massive patient assessment, stabilization, and evacuation in US history.[229] These federal NDMS DMAT and related federal staff assets, though normally under the command of State public health, emergency management and FEMA, to a great extent deployed and acted independently, albeit with FEMA approval (required for eventual reimbursement under the Stafford Act). They secured helicopters for hospital personnel evacuations, staffed emergency clinics, and moved to where they felt they were needed. They worked under life safety provisions (serious or grave conditions that required immediate medical responses outside of the accepted command and control procedures, in order to forestall patient mortality and morbidity). They were not challenged by upper levels of management in FEMA or any of the other ESFs. These self-deployments demonstrate that Emergency Managers have wide latitude to use their individual judgment under life safety authorities when efficient organizational structures do not exist, though these can sometimes be career-threatening initiatives, especially since a two-week deployment of a DMAT element can easily approach seven figures. Unfortunately, too often HHS field staff (ASPR Regional Administrators, for example) attempt to declare too many of their requests for reimbursable assets as required under "life safety." Occasionally feeling one's mission is more important than it actually is, is understandable in the midst of a disaster, but nonetheless can complicate required missions under ESF8, most of which are definitely not first response, life safety type missions. None the less, HHS's NDMS has had a history of quickly amassing and deploying assets early in the disaster cycle.

If we can include Department of Defense assets in the overall federal response, Lieutenant General Russel Honoré took charge, leading federal troops to help rescue thousands still stranded in New Orleans days after the storm as well as conduction a huge variety of other important missions. At its peak, the military's joint task force had 22,000 military personnel deployed, one of the largest deployments in the South since troops returned home from the Civil War.[75]

Haddow et al.[2] remind us that years of inadequate funding of federal, state, and local emergency functions left those charged with addressing Katrina often incapable of fully carrying out their missions to conduct assessments, protect the public, begin to rebuild critical public infrastructure, address serious environmental issues, evacuate survivors, care for victims and begin recovery and beginning to conduct mitigation efforts. Frequent Congressional sequesters of FEMA funding during the Obama Administration, which provides much of the flow-through to state and local emergency management functions have not been fully restored to previous levels after strong efforts to address Emergency Management shortfalls after the Katrina debacle. The failures to effectively manage and staff the many disasters major disasters in 2017, the last of which was Hurricane Maria in Puerto Rico, stand as testimonial to the almost permanently weakened status of FEMA capacities to address multiple large events.

The extended, nonfederal government and its effects on emergency management

Though permanent FEMA staffing rose to the 4000 FTE level by 2018, much emergency response authority in larger incidents has now been gathered up to the FEMA level. But the overwhelming majority of disasters are still handled by local first responders, sometimes including family members and neighbors, often with the help of local and regional agencies—such as the Red Cross, the Southern Baptist Food Service, the Salvation Army, other community organizations—and individual volunteers. It has long been understood that the first or the "Golden 24 hours" of a disaster are when the major life safety issues of survivor location, extraction, possible evacuation, and treatment occur. It is possible that some first response assets (not to include water, food, and many pharmaceuticals) that are very helpful in this Golden 24 h period have been predeployed, such as some mobile assets of the National Disaster Medical System, but that is the rare situation, and usually only at weather event such as a hurricane, that can be seen and predicted with relative accuracy. In recent years, as will be amplified later, FEMA has improved its use of predeployment staff and assets.

In large-scale disasters, local agencies are responsible for managing the event until help arrives; and for that reason, local capacity building, for example, through extensive mutual aid agreements, has been a priority for decades. Professional Emergency Managers have fought to prevent cuts in emergency management performance grants and similar programs, on which they rely to maintain and expand critical functions. Given that the decisions Emergency Managers make during response operations can have profound implications for recovery, engaging local officials and their constituencies in all phase of decision making is critical a process best started well in advance of actual emergency situations. As FEMA officers say, "Meeting your counterparts and exchanging business cards in an emergency operations center, for the first time, is not a good practice." This constant FEMA observation has been supported by research that has documented that the likelihood of effective response and recovery is increased if key personnel at all levels of government have established relationships before the disaster occurs.[2]

As previously discussed, when a disaster strikes, the first people on the scene are typically friends, and neighbors, who begin to search and rescue and provide medical assistance to the extent that they can. And another element of emergency response that has continued to change in the last decades is the contracting out of government services to private and nonprofit organizations. As a result of this, private firms, faith-based organizations, community groups, colleges/universities, and other organizations are now participating in the development of plans, provide emergency services, and even operate emergency operation centers, along with providing services and products. Many response and recovery efforts following landfall at Katrina, for example, were managed largely by contractors, raising concerns (see later) about many questionable expenditures of public monies, quality control, the use of out-of-state as opposed to local contractors, and the sheer size of the bills. The contracts cost billions of dollars, and the benefits were not always apparent.[85] Of course, contractors are private firms and participate in various aspects of the political process and can support certain candidates with contributions, but lobbying is mostly a precluded minefield for local, state, and certainly federal Emergency Managers. As all Emergency Managers know, "these situations can complicate things."

A related problem is that some contractors have large, existing contracts that are renewed and extended every year without much chance of a real review or competitive bidding; these can include contracts for federal telephone, computer, internet, and many other basic infrastructure services. High quality or not, these are a relatively "safe bet" for sometimes overworked and understaffed federal contracting review officers to use some of these existing, large contractors during disasters, particularly in the recovery period. These can be confounding situations for local or even federal Emergency Managers, but not situations that can be easily or effectively addressed in most instances. The authors have observed that federal contracting officers are often most interested in avoiding adverse contract outcomes, as opposed to seeking out contractors that may produce excellent outcomes. When time is short, safety can be the most logical outcome.

Federal, state, and local program managers have been placed in a situation in which the determination of quality outcomes grows more difficult to determine and manage as the funding, operation, and accountability of many federal responsibilities grow increasingly dispersed. Unfortunately, emergency management is a prime example. How it is that government spending continues to grow, but the number of federal employees continues to stagnate. This is not the public conception of a "bloated and growing federal government." In 2013, for example, the federal government spent 5 times what it spent in 1960, adjusted for inflation. Many new programs have been created since then, including the Department of Homeland Security (2002) and the Environmental Protection Agency (1960). And yet, at 2.2 million there were more federal bureaucrats under Ronald Reagan, in 1984, than when Barack Obama was President (about 2 million). The Defense Department has 800,000 civilian workers, but almost 700,000 contractors. Homeland Security has more contract employees (200,000) than permanent federal staff, at roughly 190,000. The nonprofit sector has grown to encompass a startling 1.6 million organizations and have about $2 trillion in revenues, about one-third of that coming from the federal government. And in perhaps the most extreme example, the Energy Department spends about 90% of its annual budget on private contractors, who handle everything from radioactive waste to energy production[76] as well as searches for existing nuclear weapons.

A recent study by The National Academy of Public Administration and the Kettering Foundation (2011) concluded that programs operated by civil servants receive significantly higher scores for management and effectiveness than those run by "grant and contract-based third parties."[77]

According to John Dilullo being quoted in the *Washington Post* writing in a perspective article, "Want better smaller government? Hire another million-federal bureaucrat:"

> Yes, government is big and is dangerously debt-financed, but it is also administered by outsiders – and that is what guarantees that our big government produces bad government, too...These three players (contractors, state and local governments, and staff from non-profit organizations are often involved simultaneously in large federal programs. American's Leviathan by proxy consumes as large a share of the nation's gross domestic product (more than one third) as many supposedly more socialist European governments. *(The Washington Post, John DiIullio, 29.8.2014)*

The trend toward the continuing huge and likely growing presence of nonfederal staff in emergency management is not likely to change. In short, the new context for emergency management is one in which the Emergency Manager must work closely with social, political, and economic networks whose trust must be earned, and over which the Emergency Manager has little or no authority; it is also a context in which state and federal officials have limited authority, and therefore, also cannot assume that they can try to preempt local prerogative. This new networked environment requires greater investment in relationship building and more inclusive approaches to decision making.[67] As previously noted, as long as private contractors can contribute to election campaigns, their numbers will grow and the number of sworn federal employees will likely continue to stagnate.

Increasing global vulnerability to disasters

Natural disasters are becoming more frequent, growing more severe, and affecting more people than ever; the reasons vary but include climate change, population growth, and shifting habitation patterns.[66] Although the number of such disasters keeps rising, far fewer people are dying from them; in 1970, 200,000 people perished annually. That figure has been dramatically reduced, thanks to safety measures such as improved buildings and flood-prevention schemes:

To reduce it still further, urban planners may have to operate on the assumption of even more extreme events. Since 1970, the number of disasters worldwide has more than quadrupled to around 400 a year. Another dataset of less serious types of weather and climate-related events, defined as causing at least one death or a set amount of monetary damage, shows an increase, too. By this measure, compiled by Munich Re, there are six times more hydrological events now than in 1980. Last year's total was the highest ever seen.[78]

Kris Teutsch[66] summarizes how the most modern disaster management strategies are being adopted worldwide, to address the problem of higher levels of natural disasters. He did not stress it, but the more effective disaster preparedness, response, recovery, and mitigation strategies have, at least in the United States, been too often been accompanied by austerity in emergency management funding:

- **Identify the disaster problems precisely**—Before organizations can improve their disaster response capabilities with new technology and training, they must have a clear idea of the problems they are trying to solve and have appropriate process in place or be developing them.
- **Overcome the problems of information across organizations by interoperability**—Information that is widely distributed and owned by different organizations must be guided by widely shared procedures and definitions.
- **Data use must be normalized**—If it is not, cross border and similar issues will frustrate quick communication, understanding, and response outcomes.
- **Manual data records must be automated**—This may be less than glamorous work, but if not done disaster response and humanitarian assistance will never be as effective as they might otherwise be.
- **Information and communications technology (ICT) standards and definitions must be shared within organizations and across borders.**

Emergency Management decision making without taking the most modern and advanced ICT seriously, will be decision making with outmoded inputs.

Many communities in the United States and across the world are becoming more and more vulnerable to disasters with so many settlements already located on waterways, near mountain passes, along coastlines, in areas prone to floods, wildfires, earthquakes, landslides, hurricanes, tsunamis, volcanic activity, and other hazards, to include rising sea levels, with all of the attendant insurance and mitigation issues. There is no need to present a long list of record breaking, climate change-related heat events, for example, suffice it to say the 2015, 2016, and 2017 have been the three hottest years in the United States since temperatures have been recorded globally. The year 2019 alone, was the hottest on record. This will affect the frequency, intensity, and interrelationships between deadly heat waves, droughts, and wildfires and human habitation increasingly located in high risk locations. In the United States, the population continues to migrate to at-risk areas, with the National Oceanic and Atmospheric Administration finding that 50% of the population in the United States lives in these areas, and that 70% will live in that zone by 2025.[79]

Continued and intensifying weather hazards related to global climate change such as the devastating Hurricane Maria in Puerto Rico which caused up to $90 Billion in damage with an estimated total of deaths to exceed 4600 (this estimate, approximately 1500 over the standard, revised estimate, will be discussed later) are becoming commonplace across the world.[80] And in the next summer of 2018 there were torrential rains that brought death to up to 100 in flooding and mudslides in Japan, wildfires engulfing parts of the suburbs of Athens, Greece, and continued record-breaking heat in 2018 in Arizona, California, and other parts of the United States. The heat and arid conditions for a sixth straight, record breaking year have helped to cause and intensify nearly continuing wildfires, becoming the new normal in California, which destroyed more acres by early July than they did all of in 2017,[81] along with the unseasonal but hugely costly ($65 Billion) and deadly (185) Superstorm Sandy. But the consistent growth of serious disasters does not necessarily educate the public and maintain its focus:

> However, the combination of imagined disasters and real ones—from the Northridge, California, earthquake of 1994 to the terrorist attacks of 2001, the Asian tsunami of 2004, and the hurricanes of 2005, does focus public attention to prepare for disasters. But whether greater public awareness translates into increased understanding of disaster and greater willingness to prepare of it is questionable.[67]

As always, the wise and studied Emergency Manager has a good knowledge of what can happen, and of how to plan for, respond to, recover from, and mitigate the events. The difficulty is that the public is all too ready to forget the last serious event as time passes and is not anxious to invest the resources in preparing for these as well as the more frequent and mundane events. Unfortunately, this coincides with a decline in the public's confidence in the ability of the government to protect it.[67] Sometimes, austerity situations can actually assist emergency management as has the intensification of mutual aid among US communities, large and small. What often started as a situation intensified by local scarcity, the Emergency Management Assistance Compacts (EMAC) in the United States have improved and coordinated preparedness and response efforts across the United States, where not just assets but technology and communications capacities can be shared and mutually enhanced. But progress, even in the most advanced nations, is not always uniform. For example, a recent study of natural hazards and spatial planning in selected

European Union members has shown that spatial planning is only one of many factors in risk management and that it is, in general, not involved in risk assessment. Further, multirisk assessment approaches are not used in planning practice, risk indicators are hardly used, and vulnerability indicators are not used at all when planning practices in Finland, German, Greece, Italy, Poland, Spain, and the UK have been studied.[82] And this despite the more frequently highly socially developed traditions in the various EU countries.

The professionalization of emergency management

Until recent decades, most Emergency Managers found their way into the "business" more or less by chance, often from earlier time spent in the military, rather than by preparing themselves specifically for their vocation; this has changed in this last decade or more, with hundreds of colleges and universities in the United States and around the world offering degrees and certificates in emergency management, national security, and related fields. This will mean that the new generation of Emergency Managers will be able to take advantage of emerging technologies as well as enter their chosen field very well prepared.[2] All of this is appropriate because the job, though not necessarily the required resources, is much more complicated than it was in 1993, during the first attack on the World Trade Center, during the 911 terrorist attack, as well as the failed responses to Hurricanes Katrina and Rita in the 2004, 2005 periods.

In addition to being well trained and well educated in emergency management, effective Emergency Managers are more than ever required to be skilled communicators, "big picture thinkers," and adept at administration and politics. Craig Fugate, Emergency Management Director for the state of Florida (and later, FEMA Director) has argued, for example, that although the Incident Command System (ICS) is useful for structuring response efforts, "…it is only a tool and may need to be adapted to circumstances….also observed that "ICS zealots" can actually hamper response operation by limiting flexibility.[67] Fugate has also been an early and strong advocate of local "resilience," regardless of federal support levels.

Increasingly, Emergency Managers have to "Play in other sand boxes" such as those related to zoning, insurance, land use, and other aspects of local government. To succeed, Emergency Managers need to have developed good interpersonal skills, good political skills (or at least to have access to those endowed with those skills), good administrative skills, as well as skills in emergency planning, coordination, and other technical functions.[84] And, despite the growing need for talented and sophisticated Emergency Managers, many Emergency Management offices are still grossly understaffed and underfunded, with too many communities not even achieving basic emergency management capabilities, plans, communications capabilities, and equipment. Homeland security plans that do not include an all-hazards perspective that cannot be efficiently adapted to nonterrorist hazards and federally sanctioned decisions that discourage information sharing within and between agencies will continue to be a source of conflict in the field. Among the most important tasks for the Emergency Manager in the coming years will be to reconcile the priorities of local emergency management with those of the federal government amidst relatively austerity conditions (Ref. 67, p. 21, 22). One of the major new "sandboxes" will be the newer Automated Data world. For example, though "big data" may just slow things down in the response phase of a disaster, much more intense data can certainly assist recovery and mitigation options for the Emergency Manager.[86]

Dun and Bradstreet has used its massive business database working with FEMA to assist businesses to recover after major disasters. This interfaces well with the use of favorable, low interest Small Business Administration loans available to help business recover and even develop. These are the types of newer data use options disregarded by the Emergency Manager at the risk of his or her long-term effectiveness.

The scope of education and training now available to Emergency Managers is huge, comprehensive, and constantly expanding. For example, many states such as Indiana, Texas, and Georgia come to mind, as conducting extensive emergency management and first responder training. But it is FEMA's Emergency Management Institute that is still the focal point of training for the nation. Each year up to 2 million students attend various courses held by EMI and the many institutions it works with. At any given time up to 550 attend in-person training at the EMI's beautiful Emmitsburg, Maryland campus, another 100,000 attend EMI training across the nations, and 150,000 participate in EMI exercises. And, of course the International Association of Emergency Managers offers the Certified Emergency Management Certificate (CEM) valuable as a credential, as well as for the basic knowledge sets it imparts.[87]

The strong trend toward higher levels of training, education, and credentialing of Emergency Managers is mirrored in the profession's more effective uses of data, social and mass media, local partnerships with private business, the faith community, and nonprofit institutions. This comprehensive type of thinking is reflected in the recent development of a new app that helps South Carolinians stay safe during hurricanes and other natural disasters. The App includes GPS assistance, evacuation advice, insurance information, emergency meeting places, and much more.[88] Such an initiative that merges so many fields of expertise would have been unheard of a decade or more ago.

Although the Midwest farming crisis that is accompanying global climate change is relatively slow moving, and not a subject of immediate headlines, the related economic, political, and farming issues will need to be continually mastered and updated by Midwest Emergency Managers, who have little if any authority regarding these issues, but who will have to deal with their many effects. Even a summary reading of the *2018 National Oceanic and Atmospheric Administration's* 1600 page Climate Report[89] brings out that the Midwest farming areas, more than any other region in the United States will increasingly be challenged economically by warmer, wetter, and more humid conditions leading to a greater incidence of crop disease and more pests that will diminish the quality of stored grain, greater incidences of crop wilting, and extreme threats posed to livestock. During growing seasons, temperatures are projected to climb more in the Midwest than in any other region. Without technological advances in agriculture, which will severely reduce the farmer's bottom lines, the onslaught of high-rain events and higher temperatures will reduce the Midwestern agricultural productivity to levels last seen in the downturn of the 1980s. As the economy shrinks, and international exports shrinks, Midwestern Emergency Managers will be faced with shrinking resources as the local area's ecologies grow increasingly as fragile as the damaged farming sector.[90]

Unfortunately, the lack of systematic and realistically funded recovery and mitigation, particularly as it affects the poor and minority victims of disasters, can too easily place the Emergency Manager in the roles of social worker and community organizations specialist, roles he or she is not trained for or well-resourced in. These thorny issues will be dealt with in sections later, as well as in later chapters, but not in the depth that they might otherwise deserve.

The poor and the vulnerable as a special concern in emergency management

It has long been known that a strong correlation exists between those most affected by disasters and poverty. Because of several factors, including the inability to afford preparedness and mitigation measures, the low rental and purchase costs associated with high-risk land and a general lack of knowledge concerning risk and its sources, the poor are more vulnerable to disasters and therefore find themselves repeatedly subject to them. While this is much more apparent in the developing countries, where the bulk of annual disaster deaths occur, risk factors based on poverty and social conditions also exist within more affluent countries. A recent review of the minimal housing aid provided by FEMA to the many poor residents of Puerto Rico attest to this fact.[91] Financial status deeply affects a population's and its members' individual abilities to achieve protection from the

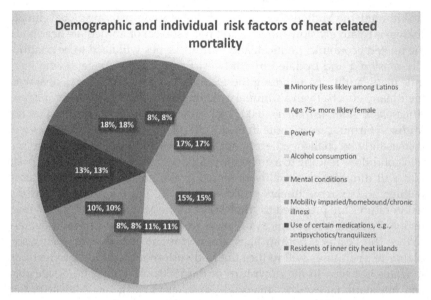

FIG. 3   Emergency managers must look for different mitigation strategies based upon a variety of factors.

consequences of disaster. Financial well-being, however, does not indicate that an individual or society *will* protect themselves; rather, it is just a measure of their ability to do so (Fig. 3).

There are other factors that may also be learned from the economic profiles. For instance, the poor are often marginalized and forced to live on more dangerous land. Their housing is more likely to be constructed of materials that are unable to withstand environmental pressures. They are more likely to have zero tolerance to delays in basic necessities that often follow disasters (Fig. 4).[2]

As demonstrated in the roles of those dying in urban heat islands (discussed in more detail below), and as Hurricane Katrina demonstrated, the poor, the elderly, and those afflicted with chronic conditions need extraordinary assistance, and are often hardest hit by any disaster. In New Orleans in 2005 many residents were unable to evacuate because they lacked the necessary transportation, financial resources, and medical assistance or chose not to evacuate because they did not understand or believe the warnings, needed to care for housebound relatives or wanted to protect their pets or belongings from the obvious dangers of the neighborhoods.[67] In short, the people on the roofs of flooded houses did not stay there by choice, they needed help. Similarly, of the many failures of FEMA's response and recovery missions in Hurricane Maria in Puerto Rico in 2018, was to not immediately begin hiring people "to clean up," to drive supplies, and so on, so as to inject money into the shrunken Island economy,

Urban heat Island

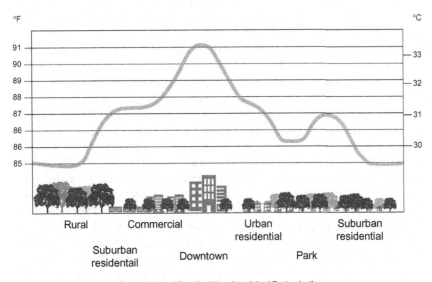

Source: Adapted from the Urban heat Island Basics, by the
Environmental Protection Agency, 2019

FIG. 4   A forecast for 95 with 90% humidity is not an equal threat.

helping people who had just lost their jobs and whatever modest incomes they had, and were already in economic need. In upper Minnesota late winter flooding a decade and a half ago, many houses were valued in the $35,000 range but were often totally destroyed or would require repairs that exceeded the value of the house minus any reimbursement FEMA provided. Virtually none had FEMA or private flood insurance. The real difficulty was that even with low interest, long-term loans from the Small Business Administration, those who were lucky enough to have Social Security had the lowest amount and were not really able to service even these loans. The wise Emergency Manager understands that at least as far as disasters go, the poor, many of the elderly and those otherwise afflicted by life altering circumstances must be addressed with the needs they present. Securing the trust of these individuals is a necessary part of preparedness in many big city environments. Unfortunately, on occasion, the urban poor, often members of minorities, are singled out for not being able to care for themselves as a group, as often occurs in rural America. But the comparisons are far from fair.

All of the above demonstrates that the victims, particularly if they are members of minority or in poverty, having survived the initial disaster find themselves falling through the recovery cracks; they may not be covered by insurance or eligible for federally backed loans, even if they could afford to pay them back and the support FEMA can provide may be too limited. Many residents of Houston, victims of Hurricane Harvey, still found themselves in temporary housing a year or more later. The Emergency Manager can easily be placed in the role of what appears to be a recovery specialist with no firm assets to distribute, dabbling in social work, a field in which he or she is neither trained nor resourced in. "As a country, we're now getting pretty good at immediate relief response," said Bob Ottenhoff, president and CEO of the Center for Disaster Philanthropy. "FEMA has a role, the big nonprofits like Red Cross and Salvation Army have roles, and there are many faith-based organizations that get involve. But long-term recovery is not as well-coordinated, in part because the funds available are really kind of a crazy quilt of differences." The US Department of Housing and Urban Development provides assistance through its many programs and there are philanthropic sources— such as nonprofits, churches, businesses, individuals—that provide relief, but navigating the process of getting that assistance can be a challenge. Ottenhoff said that often the survivors just lack the wherewithal to navigate the system. His organizations try to help by aiding the affected communities in setting up long-term recovery committees, which identify local people needing help one at a time, but the challenge is huge.[92]

And, if it is not enough that communities hit by disasters have to go about rebuilding under the worst of circumstances, "storm chasing" scam artists can victimize the poor again, this time with false promises of help and overpriced and inefficient services, if provided at all.[93] Well attuned Emergency Managers need to constantly be aware of these types of criminals.

Private business approaches to disaster response communication with AT&T

In 2017 as Hurricane Maria left Puerto Rico it left 85% of the Islands cell towers down, as well as the majority of internet and telephone lines. AT&T was on the ground just as the storm was leaving. It used a variety of strategies to restore communications in parts of Puerto Rico including its helicopter Flying Cell on Wings (COW) which helped provide data, text services. The COW hovers 200 to 400 ft above the ground and covers 40-square miles enabling Emergency Operations to proceed more smoothly. In previous disasters AT&T also used Cell on Light Trucks (COLT) which are high-capacity antennas equipped with high-capacity antennas. It is difficult to place too much stress on communications during and after a serious disaster, preparedness activities with such private organizations as AT&T can be critical.[94] More information is available at att. com/public sector.

However, as advanced as newer communications systems are, the CDC in its *Risk + Emergency Risk Communications* handbook stresses that during a disaster governmental communication must be "Be First, Be Right, Be Credible," if they are to have a positive effect on the population that is being addressed. Many studying the terrible "Spanish Flu" of 1918 point out that communications were not always accurate or believable and did not have as much positive effect as they might otherwise have had, had they been so.

> Much federal waste and duplication in our federal government is caused by mass outsourcing of program implementation to state and local governments, private non-profit entities such as universities and for-profit companies.

Though this topic will be covered in more depth in other chapters, it is important for Emergency Managers to begin considering why there have so many highly publicized failures in the federal government in recent years (the response to Hurricane Katrina, the many VA scandals, the failures during Hurricane Maria in Puerto in 2017, etc.). Congress has habitually chosen the medium of grants and contracts not so much because it loves the states more but because it loves the federal bureaucracy less. Congress loves action but it mostly hates bureaucracy and taxes, which are the instruments of action. Overwhelmingly, it has resolved this dilemma by turning over the bulk or administration to state and local governments or any organizational instrumentality it can lay its hands on whose employees are not counted on the federal payroll.[190] Emergency Managers are immersed in this outsourced world and need to face the situation with knowledge and caution because it is unlikely to change anytime soon, regardless of the President or political party.[191]

Failing infrastructure must be considered in many disaster situations

Even though it has gotten to be almost a media cliché, the reality is that during these last decades the nation's infrastructure has been consistently falling into a state of disrepair that will be made much worse by almost any weather-related disaster. There are political, economic, and even ideological reasons for this sad reality, but dealing with them is certainly beyond the scope of this book. The obvious weakness of many roads, bridges, sanitation, and sewage systems is evident to the eye, anywhere across the nation. Often, these gradually failing infrastructure components can be seen as contributing to the gravity of naturally occurring disasters. Edward P. Richards, the Director of the LSU Law Center Climate Change Law and Policy Project states some of the problem very concisely,

"A changing climate, with its increased river runoff, more severe precipitation events and sea level rise, will make it crucial to strengthen these dams and related infrastructure such as levees, reservoirs, floodgates and storm drains."

Most Emergency Managers are aware of the 16 critical infrastructure sectors that have been listed by the US Department of Homeland Security.[96] Though there is no pool of available resources to address the serious infrastructure dilapidation the nation faces, nor are there pools of available resources to address the effects of crumbling infrastructure in weather-related disasters, the state and local Emergency Managers must locate, map, and plan for not just the failure of the key infrastructural components, but how these will be affected in any serious naturally occurring disaster or in any induced event categorized as terrorist inspired; this is more than enough work for any Emergency Manager, even if there were no other responsibilities. Since many of these components are privately held, there are good opportunities for partnering. For those that are publicly held, public private coalitions may be the only way to develop realistic preparedness plans as it is not likely that taxes will be raised, or existing public revenue streams will be allocated. The Emergency Manager must outline the problems, calmly state the difficulties, and do the best that he or she can.

To give just one example, though a particularly hard one to deal with, consider the Island of Manhattan, and the vast pool of finance, commercial and communications networks that are focused there. With rising seas, there will likely be an eventual requirement to separate most of Manhattan from the sea by a series of walls, levees, planned flood lands, pumping stations, and much more. Consider that each of these components will be a terrorist target of high interest and will require huge security expenditures in addition to the tens or hundreds of billions required for the main wall projects. By 2018 the City of Miami has already begun planning to spend what will no doubt eventually be billions in significant efforts to fight flooding. There are many other examples. The Emergency Manager may need to call in "outside reinforcements" such a student interns, university planning or emergency management departments, or similar intellectual assets to lend a hand.

Weather-related hazards, technological hazards, and induced events (terrorists' use of weapons of mass destruction)

The next subsections will address decision making in various on-the-ground emergencies, disasters, and catastrophic events. There is no attempt to describe specific actions to be taken in specific preparedness, response, recovery, and mitigation events, because comprehensive "How to" materials exist, for example in (1) Haddow et al.'s *ICMA Introduction to Emergency Management*, (2) in Draebeck and Hoetmer's original ICMA "Green Book," 1991 in (3) ICMA's *Emergency Management: Principles and Practice for Local Government*, by Waugh and Tierney, 2014 or in (4) Fagel's *Principles of Emergency Management and Emergency Operations Centers (EOC)*, or in the many more functionally oriented texts such as (5) Henderson et al.'s *Bioterrorism: Guidelines for Medical and Public Health Management*, JAMA, 2002, or (6) Glarum et al.'s *Hospital Response Teams*. Instead, select aspects of decision making in these circumstance will be addressed, with an emphasis on response decision making, as opposed to the slower moving, longer term processes of preparedness, recovery, and mitigation, though these will receive coverage in select and more critical areas: (1) Weather-related hazards will include floods, earthquakes, tsunamis, hurricanes/winter storms, tornadoes, draughts, heat waves, wildfires, and non-terrorist- related technological hazards. (2) Terrorist events, especially with the strong emphasis US Department of Homeland Security puts on them, especially their enforcement aspects, will receive only secondary coverage here, as Emergency Managers do not play as central a role in decision making as they do in naturally occurring and non-terrorism-related technological hazards, and are in more of a supportive role regarding supplying commodities, equipment, and staffing. In short, injured residents and damaged infrastructure, whether resulting from weather related events or terrorists events, must still be deal with by using existing resources under the all-hazards approach.

Continued urbanization across the world

By 2018 approximately three in five cities (679 of 1146) worldwide with at least 500,000 inhabitants are at a high risk of a natural disaster according to the United Nations in its *World's Cities in 2018* report.[100] They were vulnerable to cyclones, floods, droughts, earthquakes, landslides, or volcanic eruptions—or some combinations of these. Megacities of more than 10 million are most exposed, with only three of them (Moscow in Russia, Cairo in Egypt, and Kinshasa of the Democratic Republic of the Congo) deemed to be at low or no risk for the six disasters analyzed by the study. And the death toll has been severe with 1.3 million dying in the last 20 years and leaving an astounding 4.4 billion injured, homeless, or in need of emergency assistance.

Sadly, the poorest of the cities are often most exposed. With some large cities, such as Naples, Italy exposed to a serious volcano potential, but fortunate to be in Italy with an otherwise excellent emergency response capacity. Recall, the same Mt. Etna that nearly completely destroyed ancient

Pompeii two thousand years ago is still active and still a constant threat. Or Houston, Texas, exposed to serious hurricane and flooding threats but also fortunate enough to be in Texas, a US State with excellent first response emergency management assets, whose deficiencies are in the longer term recovery, reimbursement, and housing situations that would otherwise be addressed by effective national statutes and funding at the FEMA level, outcomes that are far from assured. As the world continues to face the continued and intensifying threat of global climate change and increasing overpopulation, there are those who are studying newer methods of city growth and development that can mitigate the effects of many of the natural disasters they are increasingly faced with. The difficulty is the same lack of enlightened political and governmental leadership that is mostly unable to stop this festering urban situation. It is also unlikely to have the determination and the ideas to implement some of these forward-thinking urban solutions.

The local Real Estate Board in the US is an institution that has frequently been categorized as the most knowledgeable group in any area regarding the composition, direction, and the future of urban structure and neighborhoods. Their influence is consistently strong in the real estate area, they lobby city councils and state legislators and they work mostly behind the scenes, known mainly to each other and the larger real estate, building, and banking interests. The Emergency Manager needs to be aware of their power and also of the fact that they can be powerful supporters or enemies of many aspects of mitigation. But, since making them supporters of wise mitigation may be a mission mostly out of the realm of possibility for most Emergency Managers, as well as most non-Real Estate leaders. However, since it is always better to know more than little, the Emergency Manager can have an insight into whether powerful interests are entering a neighborhood or urban area, or are leaving things pretty much as they are. Call this type of local information, "predisaster intell." If the urban Emergency Manager is lucky, this type of insight can be quickly learned and filed away, and never used. If some neighborhoods are affected by serious disaster conditions, knowing who owns what and who was buying or redeveloping what is never a bad thing. Sometimes a trusted real estate agent can supply more than enough information for the Emergency Manager. Most real estate and urban planning students are advised to drive around city areas frequently, doing so-called windshield surveys to develop a constant feel for changes (not a bad practice for Emergency Managers during and after disasters—of course to the extent practical, or even legal, by using a drone if driving is not possible). Big realty companies usually place little metal signs of the same color, in the same location in properties they own. When the signs are shiny and new, it is a good indication that the building has been recently purchased, conversely, when most signs in the neighborhood are dull or even rusty, it is a sign that ownership patterns are relatively stable. Regardless, urban recovery and mitigation strategies are important, involve many powerful actors, and are complex. But experienced Emergency Managers already know most of this.

Almost 50 years ago far thinking urban architect and scientist Paolo Soleri developed the concept of the "arcology" merging the terms "architecture and ecology" to develop the concept of huge, multipurpose buildings that minimized the impact on human structures on overall ecology. In other words, the buildings featured living spaces, work spaces, recreation, plants, and natural surfaces so as to reduce the human footprint on the earth's surface. In a recent piece on modern concepts of city development facing the challenges of global climate change,[101] among the far-thinking ideas are buildings that are huge, tall Vertical Forests, buildings designed by Stefani Boeri that provide living space as well as tree and vegetation that remove carbon dioxide from the air. In one of his models in Milano, Italy he has developed a pair of apartment buildings that feature 20,000 plants and 20 species of birds. The difficulty with these concepts is that they are only likely in rich, advanced countries where they are certainly needed, but not in the poor, overpopulated, disaster prone megacities where they are really needed, but impede from experiments because of political and fiscal restraints. Of course, these are the far edge of urban thought, but as time passes, if Col. Boyd is to be our guide, a healthy appreciation of even the most advanced thinking is a good thing for the Emergency Manager to have at least somewhat cultivated.

Dr. Lucy Jones, from her long experience in disaster response, planning, and research offers a few suggestions toward making cities more resilient regarding avoiding some of the worst aspects of natural disasters:

- **Educate yourself**. Find and log your risks. Prioritize potential responses. Stress what is preventable.
- **Engage with local leaders particularly elected officials**. For example, if you think that building codes need improvement, involve the representatives.
- **Work with your community**. De. Carvallo inspired the king and his subjects to start rebuilding in Lisbon, before grief and despair took over. In Japan, Maki Sahara moved beyond her life as a housewife to help mothers of Fukushima cope with their fears of radiation.
- **Remember that disasters are more than the moment at which they happen.**
- **Think for yourself.** Don't rely on existing solutions and infrastructure to be adequate if severely tested. It was over reliance and overconfidence that led Otsuchi's city council to meet behind a sea wall leaving them in the path of a deadly tsunami.

Weather-related hazards

Floods

Floods are the most common type of disaster worldwide, accounting for 49% of all disasters from 1900 to 2013. During this time period China has experienced 9 of the 10 deadliest floods losing 3.7 million in 1931, 2.0 million in 1959, and 0.5 million in 1939 accounting for 5% of all disaster-related deaths worldwide. US flood losses have been significant, though are nowhere at the levels China and India, for example, have experienced in recent decades. Direct economic losses from the 1993 Great Midwestern Floods exceeded $10 Billion while flooding after Hurricane Katrina was estimated at $25 Billion. US trends toward increasing population density near coasts and in floodplains suggest, along with continued and intensifying global climate change, even higher levels of future flood-related damage. In the US up to 20,000 communities are subject to a substantial flooding risk with 7% of the nation's land area being subject to flooding. Again, in the US, as much as 90% of the damage from disasters is from flooding, including flooding related to hurricanes and storm surge.[102] With some exceptions, mitigation efforts are difficult and uncertain, and response efforts, though easily seen developing, can be complex, involving many groups and assets for which the Emergency Managers does not have firm control. Across the world, most Emergency Managers are virtually always mindful of potential flooding in their own or adjacent areas and are aware of the potential for death and injuries, the economic losses, and even the potentially huge public health impacts presented by floods. In the US Emergency Managers with flood prone areas will need to remain vigilant regarding public health impacts of flooding to include:

- Loss of potable water and adequate sanitation (easily lost with frequently river-based plants becoming inundated),
- Crop losses and disruption of food distribution
- Loss of shelter and population displacement (with shelter staffing, sanitation, safety, and related logistics issues as recurrent problem areas).
- Toxic Mold Exposure (the development of various types of mold on wall board and wood that has been soaked from contaminated flood waters is one of the more costly items that accompany rebuilding after floods; too often cleanup workers are not fitted for appropriate respirators and can easily be exposed to health threatening molds).
- Disruptions of health care services (The Emergency Manager must be vigilant that the uncompensated medical services provided by the NDMS Disaster Medical Assistance Teams don't stay in the area too long, thereby damaging the medical businesses of local providers). After a week or two external uncompensated medical services need

to be reviewed regarding a specific phase out or termination date. Leaving an area with a shrunken local primary medical care system after a disaster recovery is a situation that needs to be avoided).

* Disruption of Public Services (power outages, mail service, solid waste management removal activities are often stopped affecting businesses as well as citizen's normal lives).
* Flood-Related Myths (Public myths regarding the need of large-scale public health interventions related to communicable disease presentation and control are common after flooding disasters. Even after learning that there is little scientific basis for starting such programs, intense pressure on Emergency Managers often persists, tempting them to do something, just to be viewed as having responded, even though there is little or nothing to respond to in most cases.).
* Flood-Related Morbidity (Poverty is a key risk factor during and after floods with low-income residents typically living in areas more likely to be damaged). Diseases normally occurring in an area are usually intensified during and after floods. In the first 6 weeks after a flood, higher income residents experience half of their injuries during the cleanup with the other half related to illnesses related to the flood. Diarrheal illnesses, increasing vector borne diseases from the much higher levels of mosquito infestations can be expected as well as dermatological issues such as rashes, and even Methicillin-resistant *Staphylococcus aureus* (MRSA) can occur more frequently along with injuries from the flood, more likely related to the cleanup.
* The Social Flood Vulnerability Index (The Emergency Manager needs to be mindful of issues that make citizens more prone to injury during the flood and that can hamper recovery).
  o Unemployment
  o Overcrowding living conditions
  o Nonownership of a car
  o Nonownership of a home
  o Long-term illness
  o Single-parent family
  o Elderly.[102]

The US has been fortunate to the extent that flooding and related conditions have not killed or injured residents to the extent they have in many other countries, though Katrina's 2000 dead were mostly related to drowning. But flooding will continue to be a growing hazard because of the continued popularity of coastal and floodplain development and of the intensifying of weather patterns that have accompanied global climate change. According to the US National Oceanographic and Atmospheric Administration (NOAA)[103]in most years, flooding causes more deaths and damage than any other hydrometeorological phenomena. In many years it is common for three-quarters of all federally declared disasters to be due, at least in part, to flooding and in the US, floods cause as much as 90% of the damage from disasters (excluding draughts).

A brief look at the 2018 flooding in North Carolina demonstrates how flooding issues can build on each other, finally presenting an almost insurmountable problem. The following year, for example, saw approximately 1/2 million acres under water in the Mississippi Delta region as well as record flooding on the Mississippi River itself. In North Carolina the initial flood water swamped coals ash ponds at power plants while rising waters engulfed private septic systems in backyards. This toxic mix inundated huge hog waste lagoons on farms. Then the torrent overwhelmed municipal wastewater treatments plants in large and small towns. Now flood water was not only toxic and murky but was a problem that dwarfed all others for a time complicating transportation, rescue, and existing public health issues. An estimated 121 million gallons of untreated and partially treated sewage washed out at more than 200 wastewater treatment systems intensified by existing industrial pollution. The bilge was disgorged in nearly 600 incidents. And people get sick from it all.[104]

In recent years Europe suffered over 100 major damaging floods, including the catastrophic floods along the Danube and Elbe Rivers in 2002. Since 1998, floods have caused some 700 fatalities, displaced about half a million and experienced 25 billion EUOs in insurance losses. Because at least 10 million Europeans live in the areas of extreme risk, such as along the Rhine, with 165 Billion EUR at risk and growing concern for related public health issues, the European Union began developing a Flood Directive in 2005. The Flood Directives address river basin and coastal mitigation objectives for human health, the environment, economic activities, or quality of life issues affected by floods. The Directives include enhanced communication systems, information exchange and dissemination, coordinated research with stronger links between the research community, the authorities responsible for water management and flood protection, and improving coordination between relevant community policies. The purpose of the Directive will be to require that Member States manage risks of floods to people, property, and the environment, develop flood mapping and risk management plans with EU funding, and to coordinate with relevant Community policies.[105]

Unfortunately, efforts to harmonize national and regional natural hazard insurance systems as well as top-down EU initiatives are argued to be inadequate, to date. The overarching question being addressed is should the EU aim at a unified hazard insurance, since so many countries differ in their approach. Switzerland, for example, has many positive changes underway, while other countries such as German, Italy, and the Netherland lag far behind.[106]

Even the most talented and socially aware Emergency Manager serving at local, state, or federal levels during a catastrophic flood cannot single-handedly possess all of the FEMA/regulatory, legal, hydrologic, real estate, environmental, public health and medical, social and demographic, fiscal, and so on, information, data, perspectives, and networks required for an effective flood response and recovery, not to mention the difficulties associated with mitigation. And recall Herbert Simon's warning, from earlier chapters, that we as humans have clear limits to what we can understand and process, especially in a short period of time and under significant duress. This does not mean that the normal federal, state, and local processes are inefficient or should be disregarded, but that these administrative and regulatory structures are constantly changing, and sometimes require quite a bit of time and resources to address. And that too often meeting the needs of complex processes appear to be as important as the response itself. Of course the higher in the management chain the Emergency Manager is, the easier this access will be. Based on this the Emergency Manager will need direct access to at least one individual expert in the following areas:

(1) The latest federal, state, and local procedures and regulations. And, as everything written here demonstrates, the federal and state rules, funding, and procedures will continue to change, usually without adequate funding, regulatory, and staffing support.
(2) At least some staff who can access, summarize, and disseminate data to meet the growing and incessant political demand for "Data and Information" to report out.
(3) Recent ground truth relating to each of the most significant ESFs in play.

Coastal flooding

There is a final area related to flooding that a growing number of Emergency Managers need to become more knowledgeable about (though certainly not to manage)—coastal flooding, along with the maze of FEMA flood insurance, private insurance, mitigation, and rebuilding problems resulting from damage to coastal properties. These problems are worsened because, as we've reported, more people are being drawn to coastal areas despite the potential problems. And too often, their problems can be minimized by local residents, almost to an amazing degree. For example, everywhere journalist Karen MacClune went in Galveston she notices little plaques on buildings at three feet, five feet, and even 15 ft high on buildings pointing to the high water marks from previous flooding, despite the fact that an estimated 80% of buildings had neither FEMA nor private flood insurance. Hurricane Harvey caused $125 Billion in inflation-adjusted damages. Some adaptations have been made, but in areas like Houston, with its unstoppable growth, though there have been some new regulations addressing the huge flooding potential, much more needs to be done.[107] To emphasize the point that coastal flooding issues have an immediacy and a power that an Emergency Manager with at risk locations should not disregard, coastal property values in the Miami-Dade County coastal areas have already lost an estimated $465 million between 2005 and 2016, with the totals in the last 2 years no doubt driving the losses higher.[108]

As global climate change cause rising seas, more intense storms and storm surges, those living along the coasts are increasingly exposed. Hurricane Sandy, which experienced a generally positive FEMA/ESF response, presents an excellent and unfortunate example of flood insurance programs that too often continue to repair the same properties damaged by coastal flooding, driving up costs, and also hindering mitigation efforts designed to at least partially address the growing damage tolls. This area of coastal flooding has a level of complexity that is almost unparalleled in emergency management. It includes the involvement of powerful real estate, banking, legal, insurance (FEMA and private), environmental and homeowners' groups and their political representatives. Understanding the needs, the motivations, and the levels of power and knowledge of all of these groups in a rapidly changing regulatory, physical and political environment can almost be a full-time job. Of course, in this situation, the Emergency Manager has to know all he or she can, but it is more important to have access to trusted staff (or even nonstaff) that know more. The major players (the powerful and those less so) feel they are right, as well as righteous, and they are, at least as far as their individual self-interests or beliefs go. The Emergency Manager operates at the edge of most of the power plays, but still needs to know the rules of how these games are played.

An example of the kinds of issues Emergency Managers need knowledge of, but have very few regulatory and political tools to address to any great degree was highlighted by Edward P. Richards, JD, MPH, Director: LSU Law Center Climate Change Law and Policy Project, Clarence W. Edwards Professor of Law at the Louisiana State University Law School referencing a key article by Mark Collette ("Flood Games: How manipulation of flood insurance leads to repeat disasters"). The article appeared on a disaster law web group (disaster_law@lists.berkeley.edu), started by Dan Farber and heavily supported by legal experts such as Professor Richards. It will be summarized in some detail in order to demonstrate just some of the difficulties of the federal flood insurance programs. During 2018 FEMA decided to severely cut back its investment in the flood insurance program, and then quickly reversed itself reminding us that political decision do not necessarily have the highest outcomes in mind, but sometimes they do.

In "Flood Games..." Mark Collette, writing for the *Houston Chronicle*, observes that officials in Houston and across the nation are failing to enforce a central pillar of the taxpayer-subsidized National flood Insurance Program (by FEMA): Making sure severely damaged properties are elevated or removed from flood plains, which would appear to be a pretty straightforward mitigation method. Thousands of homes get rebuilt and then flood again, often costing more than they are worth, costing taxpayers over $1 Billion in repeat losses. The deeply indebted program was set to lapse in July of 2018 without congressional reauthorization. Under federal rules it is local officials who are supposed to assess flood damages and require demolition or elevations if the damage is estimated at 50% or more of the home's value. But telling traumatized flood victims that they will have to undertake home elevation projects is politically and emotionally hard, so officials lowball the damage estimates, putting people and homes back in the same vulnerable places. So, calculations of damage become political instead of mathematical prices, because the federal government (FEMA in this instance) puts the burden of that decision on local officials, even though the home is federally insured. The result is that everyone pays for these bad assessments. Americans pay in in the costs of disasters: in higher flood waters as a result of buildings standing where they shouldn't be, frustrating shoreline mitigation efforts; in massive infrastructure projects to protect the high-risk homes; and by becoming tenants and owners of repaired buildings in high-risk areas where developers have made a profit and then left. And, most directly, in the cost of federally subsidized flood insurance and the US Treasury's billion dollar bailouts of the program (https://www.houstonchronicle.com/business/article/Flood-Games-How-victims-local-officials-and-an-13031069.php).

After intense flooding in parts of Copenhagen in 2011, officials developed a plan for fortifying the city against flooding, especially the more unforeseen, flash flooding conditions from unexpected "cloud bursts." Officials in New York City, a partner city to Copenhagen, have investigated ways to make the city more absorbent with design changes like planting grass to replace asphalt (because asphalt does not absorb water), lowering playgrounds and basketball courts so they hold water in a storm. Then in 2012 Hurricane Sandy flooded 51 square mile, about 17% of the city's land mass. According to Lykke Leonardsen, A Copenhagen official, "The idea of creating a new infrastructure for the management of stormwater is a way of making sure that you do not experience an unwanted flood from sewer water and stormwater, because then you're not just talking about a nuisance but a health problem. Adding green space and when sewers are overwhelmed by drain water, the goal should be 'flooding by design' so that, designing landscapes so water goes where it can be stored temporarily if it cannot be absorbed into the ground."[112]

An additional point to add to the bewildering complexity of dealing with flood mitigation, redevelopment, and insurances issues, we must stress that the very damage that the flood causes lowers community wealth, may raise insurance rates and even taxes as a result. Jonathan A. Miller addresses these issues.[109] Selected and presented by Stephen Weinstein, Columbia Law Bog at disaster_law@lists.berkeley.edu.

> Federal policies and regulations with higher standards that respond to flood risk and sea level rise are being rolled back by the current administration. In that void, the threat of credit rating downgrades is expected to be a developing non-regulatory driver to future risk planning and adaptation. Several exposed communities have been downgraded due, in part, to their lost tax base from major disasters. As sea level rise manifests along the coasts, reducing property value, impacts on revenue will present new challenges in servicing debt. Credit rating agencies in the last few years have issued publications giving some notice on how climate change is to be considered in municipal credit ratings. Proactive communities, conducting planning and realizing adaptation practices in the present are likely to be spared the need to increase revenues to counter the higher borrowing costs that are coincident with a bond rating downgraded, due to likely loss of taxable properties, caused by sea level rise in the future. Municipalities that do not engage now in addressing the threats associated with climate change may have to increase taxes to offset the increased bond return demanded by investors.[110] And, if the topics of flood insurance and flood way development and mitigation are not complicated enough, the implementation of the regulations and the funding of the overall programs are subject to successive federal Administrations and Congresses.

Earthquakes

An earthquake of great magnitude is one of the most destructive natural events. In our everyday world in which the earth seems so unyielding, it is hard to imagine a force so great that it can shake the ground into standing waves several feet high; snap tree trunks in two; spill rivers and lakes over their banks; generate seismic sea waves that can race across thousands of miles of open ocean at 700 km/h (420 miles/h) and destroy virtually every structure in a city. The forces generating great earthquakes are the same forces that generate continents, and fold and break the earth's crust. When this enormous pent-up energy is suddenly released, the impact and the destructive results can be predicted—a major catastrophe.[113] In such events emergency management decisions made years before an event, in the relative quiet and safety of training and exercise events and complex efforts to spur the adoption of earthquake resistant building codes, for example, can have their impacts magnified in ways hard to imagine. Fortunately, modern seismic science continues to predict the timing of seismic events with growing accuracy, unfortunately, the progress is insufficient to provide the months (or even days or hours) of warning that would be required to more effectively address a catastrophic seismic event (Fig. 5).

Seismic Risk Map For Conterminous united states

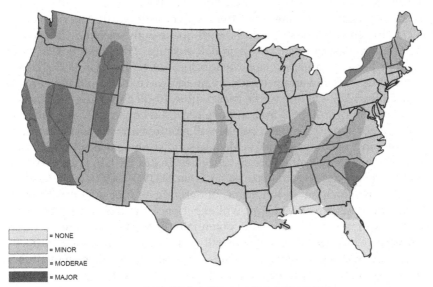

= NONE
= MINOR
= MODERAE
= MAJOR

Source: U.S. Dept. of Commerce, Geological Survey, 1982

FIG. 5   Seismic risk is a challenge as geological timing of events gives a perception of safety.

The potential extent of damage in seismically active areas can be catastrophic, with resource requirements far beyond current capacity within the US. For example, in the extensive New Madrid seismic area well over 40 million could be affected in a 7.7 event. The results of a 2008 Central United States Earthquake Consortium (CUSEC) study found that Tennessee, Arkansas, and Missouri would be affected most seriously (e.g., the fault runs right under the Mississippi River in downtown Memphis), with Illinois and Kentucky also being affected. Nearly 715,000 buildings would be damaged in the eight-state study region requiring about 42,000 search and rescue personnel working in 1500 teams. Damage to critical infrastructure (utilities, transportation, essential facilities) would be substantial in the 140 impacted counties near the rupture zone. In a serious earthquake in the New Madrid zone in the lower Midwest, the damage would be massive, including 3500 damaged bridges and nearly 425,000 households without power. Nearly 86,000 injuries and fatalities would result. Nearly 130 hospitals would be damaged, 7.2 million people would still be displaced 3 days after the event, and 2 million people would be seeking temporary shelter, requiring many primary care medical teams as well as security staff, which may be more difficult to address than medical teams. Direct economic losses were estimated at $300 billion with nearly twice this amount in indirect losses. And these figures were based on a study conducted a decade ago by CUSEC in 2009[114].

The Cascadia fault line off of the west coast of North America is predicted to generate a Magnitude 9.0 earthquake with epicenter 60 miles off the Oregon coast caused by the complete rupture of the 800-mile CSZ fault from British Columbia to Cape Mendocino, CA (Fig. 6). Ground shaking will last up to 5 min and generate a tsunami wave that could tower more than 50 ft which will impact the coast within 10–30 min of the earthquake.

It is estimated that 8.3 million people will be impacted along with 700 miles of coastline. 2.5 million people will require basic life-sustaining support within 3 h of the earthquake. Medical services and mortuary services are massive and well beyond current capacities. In just the San Francisco area, a similarly powerful earthquake could leave 250,000–400,000 homeless.[115]

In areas which have significant, underlying seismic activity, the local and state Emergency Manager will experience the difficult tradeoff between addressing the rare, and potentially huge negative effects of an earthquake balanced against the usually small chance of an event happening in the next months, or even years. With emergency management in a seemingly perpetual state of austerity, the "easy" decision of not spending much time on complex earthquake mitigation efforts (building codes, bridge safety, etc.) that require wide community involvement, time, and intense focus, and less earthquake training and preparedness activities

FIG. 6   The immediate aftermath of a Cascadia tsunami will pose unique challenges to any emergency manager. Innovative solutions will be required.

is too often a chance that usually pays off, until a time when it doesn't. And state and local Emergency Managers have the luxury of knowing that FEMA regional and headquarters offices continue to schedule a continuous battery of training and exercise earthquake events in each of the more significant seismic areas, though mitigation efforts directed at the development of earthquake proof building codes usually remain state and local responsibilities in which FEMA can be helpful, but only to a limited extent. If areas such as realistic seismic coverage in building codes, mass evacuations, and related areas focus Emergency Managers in areas mostly peripheral to these events, perhaps the Emergency Managers can focus on the development of smooth, huge, prearranged personnel and asset receipt under the various mutual aid mechanisms, interoperable communications systems, enhanced and pointed social media, and any other venues in which Emergency Managers can leverage their unique skills and contacts in the most effective manners.

A review of Building codes at State levels in the New Madrid Seismic Zone demonstrates a huge preparedness and "premitigation" effort that is yet to be satisfactorily completed. Across the United States, State and local governments have jurisdiction over adoption and enforcement of local building codes. Since developing such highly technical documents involves extensive knowledge, skills, and resources, it is most effective and practical that national model codes are adopted by state and local jurisdictions, such as the International Building Code (IBC). History has proven that communities that adopted and enforced building codes with full seismic provisions can significantly reduce loss of life and property during major earthquakes. Over two hundred years ago three great earthquakes struck within a two-month period and caused extensive damage, even changing the direction of the Mississippi River in places. There are now 9 million people living in the affected area, the New Madrid seismic area. Even though earthquakes have not occurred frequently, as in California, the area has long been recognized to be vulnerable if hit again by a great earthquake of similar magnitude. Despite the significant risk, many communities here have not taken sufficient earthquake protection, in particular adoption and enforcement of building codes and remain spotty at best. See FEMA's summary of building code status in the various states in the New Madrid Seismic Zone Area.[116] And, viewed from a national loan perspective, the losses could be huge. A recent RStreet study estimated that upwards of $200 Billion in losses to federally mortgage associations could be experienced by mortgagees who have their homes destroyed in earthquakes, but who, like most other Americans have home insurance that does not include earthquake provisions.[230]

Just over 100 years ago, for three days, San Francisco experienced catastrophic earthquakes.[117] During the period 400,000 people were displaced, fires were ignited which destroyed 500 blocks as well as 28,000 buildings and an estimated 3000 died from related causes. Much of California is in an extremely active seismic area which has frequent earthquake events. After decades of public hostility toward taller buildings, surprisingly, the city of San Francisco has been advocating a denser and more vertical downtown with more than 160 buildings taller than 240 ft. San Francisco building codes include requirements protecting schools and hospitals from major earthquakes, but not skyscrapers. A five-story building has the same requirements as a 50-story building. It would be unwise to ignore the significant seismic progress that has been made across California's many venues, for example, with freeway overpasses, bridges, and some municipal buildings having been strengthened, but nonetheless it has been calculated that one of every four buildings in the Bay area would be unusable after a magnitude 7 earthquake. A widely used welding technique was found to rupture during the 1994 Northridge earthquake (with many buildings in San Francisco and Los Angeles having not been retrofitted). But despite this, California has made significant strides in earthquake preparedness over the past century. Fortunately, many Californians live in single-family wood frame homes, which have been found to hold up relatively well during earthquakes. None of this minimizes the very real political and economic difficulties that are encountered when state and local governments attempt to change building codes, require back fitting of some existing structures, and related issues.

Useful lessons learned do not always come from the largest countries with the most advanced emergency management capacities. In February 2010 Chile experienced an enormous 8.8 magnitude earthquake, the fifth strongest ever recorded, tilting the earth's axis, by 3 in. Chile's roads, schools, hospitals, businesses, and much other infrastructure were devastated. Chile's newly elected president, Sebastián Piñera was inaugurated shortly after the earthquake, quickly mobilizing his cabinet ministers to help the injured and bury the many dead and then to repair hospitals and rebuild damaged or destroyed homes. He instructed the ministers to fully restore damaged buildings and infrastructure on a tight timeline, but he offered scant guidance on how to do so, leaving it to his trusted team to develop and execute recovery plans. The Chilean President's actions seem unusual, but his actions were quite pointed. For example, he instructed his Education Minister to arrange to get all school children back in class within 6 weeks. According to Strategy + Leadership Pinera (1) *conveyed strategic intent* in setting forth a long-term strategy for recovery that worked well within the boundaries he had to work in. (2) Pinera used *layered leadership* to ensure that managers at every level were able to implement the strategy. But mainly, as the disaster unfolded (3) the Chilean President viewed the disaster in a wide perspective but avoided micromanaging as he made the various sweeping decisions that turned out well. During the next elections he was reelected.[118]

Emergency Managers at state and local levels must individually decide what is an appropriate use of their very scarce staff and other resources regarding earthquake preparedness and "pre" mitigation in areas such as building codes and the vast community resources and interests that need to be organized successfully, a task also complicated and restrained by their relative lack of community power when dealing with major private businesses and institutions. But with the potentials to minimize the huge costs in injuries, lives, and resources even small successes can be leveraged.

Volcanoes

Volcanoes are often relegated to low risk, high damage events, and with only 2–4 annual eruptions worldwide, ending in significant mortality; they are easily lost in the procession of plan making, responses, and training/exercises for those Emergency Managers in the relatively few potentially active tornado sites in the US and the rest of the world. In the US there are at least 18 potentially dangerous volcanoes as discussed by the Nation as Geographic Magazine in an interactive website with startling videos[119] (Fig. 7).

The volcanoes were ranked based on 15 hazards including the type of volcano, the known frequency of eruptions, the threats triggering a tsunami, and potential for lahars (volcanic mudslides). The rankings also

FIG. 7   Still an active mountain range where volcanoes can't be discounted.

take into account nine exposure factors, which relate to the threats to local communities such as past fatalities and nearby populations:

1. Kilauea, Hawaii
2. Mount St. Helens, Washington
3. Mount Rainier, Washington
4. Redoubt Volcano, Alaska
5. Mount Shasta, California
6. Mount Hood, Oregon
7. Three Sisters, Oregon
8. Akutan Island, Alaska
9. Makushin Volcano, Alaska
10. Mount Spurr, Alaska
11. Lassen volcanic center, California
12. Augustine Volcano, California
13. Newberry Volcano, Oregon
14. Mount Baker, Washington
15. Glacier Peak, Washington
16. Mount Loca Hawaii
17. Crater Lake, Oregon
18. Long Valley Caldera, California

Across the rest of the world, volcanoes have the capacity to have the most extensive effects. The Laki eruption in Iceland in 1783–84 was no doubt the deadliest[120] event recorded. The overall death toll was in the millions, spanning the globe with clouds of sulfur dioxide and other deadly gases engulfing vast tracts of Europe. Blocked sunlight in Egypt caused a loss of the annual monsoons with devastating results causing huge and deadly droughts and famines. Eleven million and perhaps many more died in India and at least one million died in Japan from the resulting famines.

Volcanoes cause severe mortality or morbidity in a wide number of ways including by direct burns, suffocation from ash and other materials, trauma from ejected rocks, floods and mudflows from quickly melted snow and ice, burial under burning hot pyroclastic ash flows.

Airborne ash can affect people hundreds of miles away from eruptions and influence global climate for years afterward. Because airborne ash damages jet engines when it melts and forms gasses on the engine surface, volcanic eruptions can and do cause significant travel disruptions in recent years affecting Southeast Asia (Nov. 15, 2015 from Mount Raung, Indonesia), South Pacific (Aug. 2014, Mt. Tavurvur, Papua New Guinea), and Northern Europe (Apr. 20, from the Eyjafjallajokull volcano, Iceland). Eruptions can also trigger "lateral blasts" which can shoot large pieces of rock at very high speeds for several miles, killing by heat, impact, or burial and have sometimes knocked down entire forests.[2] Deaths can also result from hot (approximately 500°F), heavier than air, pyroclastic gas flowing down from the volcanic mountain at speeds of between fifty and 300 miles per hour.[120] Volcanic ash can contaminate water supplies, cause electrical storms, and can cause roofs to collapse under the weight of accumulated ash and other materials.

The Emergency Manager has a dual mission with impending volcanic eruptions, to speed evacuations as well as to develop mutual aid for the many medical missions that could be experienced. Rapid advances in computer simulation modeling of flow hazards and ash fallout, combined with studies of building and road vulnerability are being applied in hazard and risk formulation and are beginning to be used to support the decision making in the crises. The MESIMEX exercise in the fall of 2006 in Naples was the first "real time" exercise to test emergency plans for forecasting the direction of winds as well as the safest routes for evacuation.[121]

Tsunamis

Tsunamis are a series of wave pulses created when a large volume of ocean is displaced rapidly, usually caused by earthquakes, but also by volcanoes, or landslides. Tsunami waves can travel up to 400 miles an hour in the open ocean with a height of just a few centimeters. As they approach shallow waters near shore, they slow down but can increase to heights of 100 ft or more. Tsunamis can occur in any ocean area, but the Pacific Ocean experiences the most underground earthquakes which frequently include land masses on the edges of ocean canyons rather than on a gently sloped continental shelf (Fig. 8).

FIG. 8   Tsunami debris field, kilometers from the ocean.

Tsunamis can be intensified by the periods of high tides and when the moon is full, with its strongest pull on the oceans. Unfortunately, all tsunami early warning systems are not sensitive to volcanoes and the tsunami causing landslides they can cause. By December of 2018, Indonesia had not upgraded its tsunami early alert system. It experienced the Singkawang Typhoon struck, killing over 430. Indonesia pledged to upgrade its early warning system after this tragedy, to include underwater volcanoes and the tsunami causing landslides they could cause.

Some tsunamis can be almost unimaginable catastrophic events, for example, the world's deadliest recorded tsunami, the Boxing Day Tsunami in Indonesia in 2004, affecting 14 countries, in the northern tip of Sumatra killed almost 230,000 after a 9.1 underwater earthquake. The waves averaged between 100 and 150 ft topping at over 167 ft. Four years later, following an earthquake off the coast of Banda Arch, Indonesia a tsunami killed between 150,000 and 200,000 and injured another half million. A few years later in 2011, a massive 9.0 earthquake struck off the coast of Japan near Tohoku, reaching heights of 133 ft as it came overland. It caused a major nuclear accident, killed over 15,800 and caused over $250 Billion in damages, making it the most expensive natural disaster that ever occurred.[122]

Recent tsunamis have affected US possessions and nearby areas. At 6:48 on the morning of September 29, 2009 (17:48 UTC), an 8.1 moment magnitude earthquake struck near the Samoan Islands in the southwest Pacific Ocean. The Pacific Tsunami Warning Center (PTWC) quickly determined that the large magnitude of this earthquake, its location under the sea floor, its relatively shallow depth within the earth, and a history of tsunami-causing earthquakes in the region meant that it could have moved the seafloor and thus posed a significant tsunami risk. PTWC issued its first tsunami warning several minutes later for Samoa, American Samoa, Tonga, and other nearby island groups. In 2009 in the South Sea Islands, destructive waves only struck islands near the earthquake's epicenter, where casualties were significant. The tsunami wave reached nearly 40 ft. high in Samoa, killing 149 people there. In nearby American Samoa the waves were even higher at 23 ft, killing 34, topping 72 ft in Tonga, and killing 9 more people there. Smaller waves traveled throughout the Pacific Ocean but caused no more deaths or damage.[123] But despite the Pacific Basin's higher level of tsunami activity there have been lesser, but strong tsunamis in other places affecting the US coasts. In 1964 the West Coast's most devastating tsunami on record was generated by a deadly magnitude-9.2 quake off Alaska causing powerful waves that slammed coastal areas, including the Northern California community of Crescent City, where 11 people were killed. A surge about 20-ft high flooded nearly 30 city blocks, killed more than 100 in the tsunami zone from Alaska and down the Pacific coasts of Canada and the United States.[124] When a serious earthquake causes a serious tsunami, as it did in Indonesia in early October of 2018, the problems of earthquake liquefaction and the tsunami itself can combine to create catastrophic effects. In Palu, Indonesia, the official death total was over 2000 with 5000 still unaccounted as the worst hit areas were now being plowed after a cessation of body searches. For Indonesia, the mass burials have been recurring events, after a 9.1 earthquake and tsunami in 2004 that killed an astounding 280,000.[125]

The Emergency Manager in any coastal area, particularly on the Pacific rim, needs to be mindful of potential tsunami conditions but can only expect an event when he or she becomes aware of the underlying earthquake or ocean, bottom-altering event. Of course, public and private communications by the widest of methods are immediately required, but usually will not be as effective at forestalling deaths and injuries as one would hope. The Woods Hole Oceanographic Institute has developed a highly illustrative website, at http://bit.ly/1RWWrwwq26 to assist the Emergency Manager in virtually all aspects of tsunami prediction and response. Every Emergency Manager in an at-risk area needs to be familiar with the site and its capacities. As always, the strength of existing EMAC, JRMPO, NDMS, EPA, USAC, DCO, and FEMA (among other) relationships, and with updated contact lists, will rise to importance quickly. Also, at-risk Emergency Managers can reference the *New York Times* which has developed a "slider" graphic that allows users to see the impact of the 2011 Great East Japan Earthquake Tsunami at http://nyti.ms/1NfXnAx (Ref. 2, pp. 53–54). Descriptions of major tsunamis in the last 300 years appear on the NOA's *Science on a Sphere*.[128] The California tsunami program is developing products that will help: (1) the maritime community better understand tsunami hazards within their harbors, as well as if and where boats should go offshore to be safe, and (2) Emergency Managers develop evacuation plans for relatively small "Warning" level events where extensive evacuation is not required. Because tsunami-induced currents were responsible for most of the damage in recent events, modeled current velocity estimates should be incorporated into future forecast products from the warning centers.[129]

Hurricanes

The most extensive damage (to date) from a naturally occurring event came from Hurricane Katrina, costing approximately $80 Billion. Winter Storm/Hurricane Sandy in 2013 cost nearly $20 Billion. Storms of these magnitudes can extend their damage over many states and force recovery periods over one year (Katrina). In recent years, there have been significant advances in hurricane technology from US and European prediction models. These are disasters that an Emergency Manager can see coming, as well as the whole population, but are never really sure what the final landfall and conditions will be. Once the storm exceeds 74 mph, the Hurricane Center assigns the storm a name, state and local emergency will benefit by the increasingly accurate predictions for the storm's path after it makes landfall from the Atlantic. The storm surge from Katrina reached 28 ft, for example, and devastated communities in Alabama, Florida, Mississippi, and Louisiana. And flooding and near absolute destruction occurred across some areas in coastal Mississippi. Though wind damage, particularly with higher Category storms can be deadly, most frequently the bulk of the damage result from the flooding and potential storm surges that accompany the hurricane. So, much of the materials in the previous section of "Floods" is also appropriate to hurricane response, and need not be repeated here (Fig. 9).[2]

Emergency Managers in hurricane-prone areas quickly become expert in balancing predictive factors such as the warmth of the water in the hurricane's path, the lack or the existence of high altitude, storm killing "wind shear," its predicted paths and other items. Since the extent and variability of effects of hurricanes can be so extensive, many hurricane-prone states, such as Georgia, have found much value in hurricane mutual aid preparedness "Coalition" activities focusing on local, state, federal, private industry, and nonprofit organizations. For example, Voluntary Disaster Recovery Committees throughout Georgia are made up of local and state agencies, places of worship, and nonprofit agencies. They work with survivors on solving problems that range from financial assistance, debris cleanup, minor and major home repairs, and can include crisis and spiritual counseling. One stop shopping for body and soul.

FIG. 9   Falling trees spare no structure.

Each disaster recovery committee:

- has a mission to strengthen area-wide disaster coordination by sharing information, simplifying client access, and jointly resolving cases with unmet needs;
- helps affected families develop a plan and receive adequate assistance for recovery;
- comprises representatives from the immediate community; and
- exists with all participating organizations as equal partners.

To learn more about voluntary agencies active in disaster, more information is provided at www.ready.gov/voluntary-organizations-active-disaster.[130] Other coalition, for example, those directed at end stage renal dialysis (ESRD) users and facilities, exists across the country. ESRD groups are critical in hurricane or other catastrophic disasters that affect power supplies, as functioning, hospital-based or portable renal dialysis units (that carry their own power generators) are quite literally matters of life and death for those afflicted with serious kidney disease. For more information, see: (https://www.kcercoalition.com/en/resources/professional-resources/cms-emergency-preparedness-rule/state-healthcare-coalitions-list/healthcare-coalition-list-by-state/).[131]

A recent article in *Emergency Management* should remind coastal Emergency Managers that previously unthinkable, extremely low probability events, such as Category 6 Hurricanes, and huge, killer storm surges are increasingly possible results of continually worsening global climate change. For example, in 2015 Hurricane Patricia attained speeds of 215 miles per hour as it approached Mexico but fortunately did not make landfall at those speeds.[132] A recent study predicts that the trend toward slower moving, wetter, costlier, and bigger hurricanes that will carry more water will persist and likely intensify.[133] For Emergency Managers in Florida, in particular, this trend will require more and more preparedness as well potential mitigation efforts, at a time when overall emergency management budgets have not recovered from previous cuts, putting more pressure than ever on mutual aid coalitions and voluntary agency assistance. Emergency Managers cannot control the future, but when future trends are dire, as they are in many areas related to hurricanes, the Emergency Manager needs to try and plan and view things "two or even three steps ahead." For example, since 1970 Florida has added nearly 15 million new residents, many of them flowing to storm-prone counties bordering the Gulf or the Atlantic; the buildup is equally strong in other gulf states including Texas, which has also peppered its ocean front with development and is the reason for the huge and increasing costs of hurricanes. Homes are bigger, apartments are bigger, hotels are larger as well as the businesses servicing them, inflating costs when storms damage them. "What we build now, how we build and where we build are the three big drivers of loss. We still build too much way too close to known danger," according to Glenn McGillivray, Managing Director at the Institute for Catastrophic Loss Reduction. On a larger scale, there are continuing talks on coastal wetlands, dikes, levees, and other structures to keep the water away from the development and much more. And the costs go up, reflected in the fact that Floridians already pay the highest homeowners insurance rates in the country at the same time the national flood insurance programs looses more each year.[134]

As an example of the high-quality disaster coverage by the *New York Times*, Kendra Pierre-Louis citing many hurricane experts, summarized facts about hurricanes that are usually not known or misunderstood, even by residents in areas prone to hurricanes.

- The "cone of uncertainty" is confusing. A prime example of this perception gap is the familiar "cone of uncertainty" since in hurricane tracking. "The cone is misunderstood," said Jeff Masters, a meteorologist with the forecasting service, Weather Underground. "A lot of people look at the cone and think, "Oh, that's tee width of the storm....But no, that's where we expect the center of the storm to track." But even if the eye stays within the cone, which it does 2/3 of the time the cone can still experience catastrophic winds, floods, and storm surges.
- Water is deadlier than wind. The storm surge, the rising weather pushed ashore by the winds, is far deadlier than the wind itself, mostly because of drownings. And storm surge does not correlate with storm category and its speed can catch people off guard.
- The threat isn't limited to the coasts. The impact of the storm surge isn't necessarily the height of the surge, but how far inland it goes. Also, large storms that stall for days can dump huge amounts of water on an area far from the coast.

The scientists at the National Center for Atmospheric Research stress that better messaging before and during hurricanes is still needed. No matter how good the information is or when it is made public, many people cannot act on it for health, financial, or other reasons. No matter how much interest in getting accurate messages to those most vulnerable by the churches, synagogues, mosques, other organizations, the emergency management will have to be constantly aware of the composition of those in shelters, those who should be in shelters and those who may be barricaded at home, for many reasons.[135]

NDMS DMAT assets as a key factor in early hurricane response

Unfortunately, the otherwise comprehensive "2017 Hurricane Season FEMA After-Action Report"[136] does not mention the National Disaster Medical System. A skilled Emergency Manager finds the various intersections between the mobile, federal DMATs and the mobile, Emergency Medical System mutual aid assets available under the previously discussed Emergency Medical Assistance Compact (generally fire dept., EMS paramedics, and Hazmat assets and staff), at the various locations and levels of intensity of impending Hurricanes. Public health and medical system preparedness plans must be constantly updated to assure the adequate deployment and predeployment of assets available during the "golden first 24 hours" of a catastrophic disaster when the injured are most in need of medical care and when it can be most effective. A clear understanding of NDMS and its DMAT assets doesn't just surface from faithfully completing FEMA course requirements and even the Emergency Management Certification. It is a bit more complicated than that.

Recall, first of all, that DMATs are composed of volunteer nurses, paramedics, a pharmacist or two, a few physicians, and logistics and communications staff. Team's sizes can range from 20 or less, to 50 or more depending on the need-based configuration. Many physicians who serve on deployments will probably lose money as they usually have to pay back their practices when they are missing. Pay scales are federal, which can be acceptable for most nonphysicians. The volunteers agree to deploy for up to 2 weeks every year or so, often using vacation time from the hospitals or EMS services at which they work, when they are federalized. The training they receive is superior and most enjoy the chance to work in in the midst of serious disasters in sometimes exotic locations and under the types of austere circumstances they may have never experienced. In short, quality NDMS professionals are not "in it for the money," but rather to serve. On the other hand, there are times when teams composed of these types of professionals can be difficult to manage. As an example, a physician quitting NDMS "in a huff" appeared on a cable television show accusing her teammates of spending the day getting manicures, and implying that they were getting them at a forced discount from the local manicurist. The implication was that the DMAT staff was on duty and should have been treating patients instead of getting their nails done. The next day, on the same cable show, the team physician involved, clarified that there were no patients waiting. The issue of the fairness of the price was not brought up. Before long, critics and supporters of the NDMS began referring to it as the National Disaster Manicure System. A or so as the authors recall, year before another team physician (not leaving NDMS) complained in a mass circulation newspaper that the DMAT was wasting time at the hotel instead of being deployed, not mentioning that dispatching teams to areas of need in a timely manner was as much an art as a rigorous process. Sometimes mistakes are made and teams "waste" time in hotels, and sometimes it is necessary because of circumstances out of every one's control. One of the authors had personal knowledge of the deployment and felt that NDMS did have difficulty in deploying this particular team in a timely manner, though it was difficult to gauge whether or not a "mistake" in judgment were made. There are many other examples of the same talents and confidence that makes a successful DMAT staff can be the same traits that can make on difficult to deal with administratively (Fig. 10).

When to deactivate a DMAT is also a critical decision, and should be considered, starting the first day of a deployment. This is a key decision point because virtually all federal medical assets provide free or uncompensated services as long as they are deployed and have a tendency to want to "overstay their welcomes," as these particularly hard driving medical people might be expected to do. And too much free care can damage the practices of local physicians, sometimes convincing them to relocate their practices, not a problem local area wants to deal with along with other recovery and mitigation issues. To make the matter even more complex, most hospitals have restrictive practice requirements that preclude DMAT staff from physically working in the hospital Emergency Department, for example. Based on this, most DMAT will deploy near hospitals and will attempt to deal with the less serious, primary care cases, leaving the more complex, and more highly reimbursed cases to in-hospital practitioners. DMATs are there to assist, never to "take over" the local medical system.

Though DMAT deployments are usually worked out between state departments of public health (state emergency management agencies have varying levels of involvement), the local medical systems, the Regional Administrator for DHHS/ASPR, and the NDMS leadership in DHHS/ASPR. The wise Emergency Manager stays in the information loop, and command and control loop. And FEMA must agree to the mission

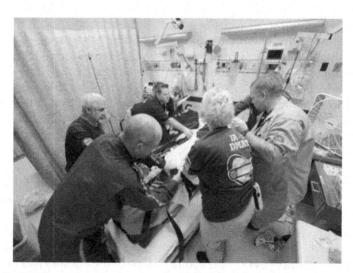

FIG. 10   DMAT members are always anxious to help manage casualties of any event.

assignment, though there is usually a lot of flexibility involved because of the life safety aspect of some, but not most missions. To this extent the local or state Emergency Manager will make security and transportation assets available if they are needed, to the extent possible, as NDMS does not carry its own security and has no air assets of its own. And sometimes, NDMS assets are used to staff the massive U.S. Comfort, a huge Department of Defense medical asset that can be of real assistance at almost any time during a disaster, assuming that there are enough patients to justify its huge cost, and that there are secure transportation methods to convey patients to the ship. Another difficulty of deploying the U.S.S. Comfort in a cost-effective manner is that there is often a tendency for upper level managers in DHHS/ASPR to "Show the flag," and to deploy the Comfort or other assets because of the drama implied even in circumstances when its use is not appropriate.

The Regional Administrators for DHHS/ASPR usually work between the state agencies, FEMA, and local areas to assure quick, safe, and effective deployment of NDMS assets. These individuals and their contact numbers should be known to effective Emergency Managers at the local, state, and federal regional levels. The sensitivity and knowledge required to balance the need to augment local emergency medical services while not keeping outside assets in local areas so long that they damage the practices of existing physicians and hospitals is one of an Emergency Manager's hardest tasks, though these decisions are generally made by ASPR Regional Administrators in conjunction with state health department and perhaps with FEMA involvement. The decision to extract a DMAT should not be made by the DMAT leadership but is best done as part of the decision tree. The reason this is so, is because the DMATs are treatment teams and not equipped or charged with medical deployment policy.

Another often overlooked asset that is deployed by the US Department of Health and Human Services is the 7th uniformed Service, the 6000 officers of the US Public Health Service, Commissioned Corps. The individuals are often confused with Navy Officers, as they are clad in uniforms similar to that worn by Naval Officers. These Officers are deployable on disasters, can work with the ASPR Regional Health Administrators, and are almost a "free asset" for states involved in disasters, "free" to the extent that they can be placed on mission assignments, and if approved, reimbursed by FEMA. They can be placed under federal, state, or local control, and are widely skilled in both medical treatment and administrations with special skill sets in disaster management, though they are more of an administrative asset as opposed to a field treatment asset, as NDMS staff is.

A final point to address regarding NDMS deployments is the conduct of a needs assessment. In past years ASPR Regional Administrators conducted Medical Needs Assessments, that coordinated with FEMA's Incident Management Assistance Teams to determine medical and public health needs, usually also involving deployed staff from the Centers for Disease Control who advise on public health issues such as potable water, disease vectors (mosquitoes, communicable disease, etc.). If there is time ASPR may predeploy staff that can conduct needs assessments or even advance NDMS DMAT elements. High schools, hotels, and so on, that meet 150 mph wind standards are frequently chosen to stage predeployed assets. Ironically, during the tragic Katrina Hurricane, NDMS had predeployed and staged team assets though security and transportation issues appeared in the general confused nature of the early days of the disaster. A wise Emergency Manager will have a state health department and an ASPR Regional Administrator contact to keep him or herself constantly grounded in the progress of ESF #8. The fact that sometimes powerful physicians accompany DMAT deployments can sometimes complicate chains of command. The needs assessment is critical for a number of reasons, but a key reason is that the Emergency Prescription Assistance Program can be opened, enabling people in disaster areas who do not have insurance to have access to necessary prescriptions. More information is available at: https://www.phe.gov/Preparedness/planning/epap/Pages/default. aspx.[137] One of the most important missions of the NDMS to the elderly and others whose prescriptions need to be met in a timely manner to prevent normally functioning people from becoming patients, or worse.

It would seem with the potential for lifesaving drama, dedicated medical staff, and interesting disaster situation, that NDMS would be a household word. Not necessarily so. Except for a negative story placed in the mass media by a disgruntled DMAT physician every few years, there are few other sources of positive public relations. Let us present a case study of sorts. During a deployment at the Republican Convention in Minnesota in 2006 a number of deployed DMAT members were having lunch at a local restaurant. Next to their table, a middle-aged man "coded," and fell out of his chair. The members got up quickly but calmly and cleared the table with their forearms, boosting the lifeless man onto the table. In much less than a minute the members determined that he had not choked and had a heart attack. Almost immediately they had his heart pumping with him breathing. 911 had already been called. The restaurant erupted in cheers just as the team members all seemed to smile shyly and go back to their breakfasts, which they had not knocked on the floor. The local Regional Administrator gathered up some phone pictures that had been taken, did some interviews, and wrote a short story to transfer to PR to work. He began contacting FEMA to get a favorable story or TV news event placement. FEMA never did place the story. The DHHS PR staff was apparently off for the weekend, so they were not able to push the story. This is not a rare type of occurrence. For reasons that are familiar, but to some extent unwise, the leadership of NDMS can too easily seek to overdeploy assets, but to keep its public visibility low, public notice that would likely boost both budgets and available assets. Depending on how it is calculated, NDMS has less than a $100 million annual budget for a nation of over 330 million.

In this same vein there have been conflicting viewpoints about the size and international deployments of NDMS DMATs, boiling just beneath the surface, for decades. As noted, the complicated, annual NDMS budget is less than $100 million covering 42 Disaster Medical Assistance Teams (as of late 2018) for a nation of approximately 330 million. There are plans to expand the existing 4600 members by approximately 2500 (although this can change quickly, and usually for reasons external to the NDMS and its ASPR leadership). And the importance of the NDMS at whatever strength level should not be minimized. For example, NDMS treated 42,000 patients and delivered 944 tons of medical supplies in the hurricanes of 2017.[138] And the NDMS is always available to assist in state and local public health and medical authorities in outbreaks (Ebola, Zika, SARS, pandemic flu, etc.).

In fairness, "routine" deployment decisions are usually made relatively quickly, meeting or even exceeding the "ideal" amount of assets. But when deployments are occurring in massive or highly visible or conflicted events the amount of powerful players can sometimes confound normal decision making channels (Fig. 11).

These issues reflect decisions are often made by powerful political appointees, and their staffs in ASPR, who make virtually all of the policy decisions affecting NDMS, though partisan issues do not seem to ever

FIG. 11 International deployments often highlight the differences in infrastructure around the globe.

had much effect on NDMS. Decisions reflect the two differing views of NDMS. One view, pretty much the existing view is that NDMS is far from a first-response asset and infrequently predeploys weather events, that exist mostly to support state and local assets, so issues like security and advanced mobility (meaning helicopters) are not significant to the mission and would absorb lots of resources (and that is correct). The other view is to see NDMS as an asset that could travel the world frequently to help many victims of disaster, representing the best aspects of America and which uses these deployments also to train and even draw an even deeper pool of volunteers (thought the present pool is superior, by most measures). Is it necessary to have this type of information in the back of one's head when pondering a potential NDMS DMAT deployment? Usually not, but as we have addressed from many different perspectives, many times, the more knowledge an Emergency Manager has about the reality and practicality of potential mission requests, the better chance there is getting more of what the Emergency Manager feels is appropriate.

Winter storms

Winter storms were relatively uncommon causes of major disasters in Canada, the UK, and the US until the mid-1990s, after which the number resulting in major disasters has since been in the double digits annually, accelerating along with the scope of climate change. Serious winter storms and their aftermaths affect both populations and societal systems, the latter of which includes the ability to respond to community emergencies, and to deliver health care outside and inside hospitals and physicians' offices. All are affected by power loss, hazardous driving conditions, and the intensification of geographic isolation.[140]

Severe winter storms can be life-changing events that isolate and disrupt families, result in closure of schools, businesses, hospitals, EMS services, and government; prevent air, ground, and water transportation; and destroy large components of agricultural, manufacturing, retail, and service industries. Public safety can be threatened when roads are impassable, security issues are intensified, power grids are nonfunctional, and telecommunications are inoperable or severely limited. Developing accurate assessments of damages in these circumstances can be complicated by the potential of a huge number of varying system failures spread across huge geographic areas with limited access to develop accurate and timely reports. With these limitations in mind, the DHHS Centers for Disease Control and Prevention (CDC) has developed rapid community needs assessment models using modified cluster sampling techniques.[233] It is just too easy to assume that helicopters will be available to do rapid flyovers providing highly useful and understandable information as soon as wind speed dips below 40 mph.

Sometimes helicopters are monopolized by political or other interests, have difficulties landing and flying in constant wind, ice storm, snow, and related conditions or have mechanical difficulties. Ice storms with impassable roads, sagging and falling power lines, dead land lines, and sporadic cell phone coverage can serve as one of an emergency manager's worst "bad dreams." Creativity, luck, and persistence will remain major weapons that are always available to address some of the more serious communications-related aspects of severe winter storms. And, as noted previously, the growing availability of highly effective drones is increasing quickly.

In mid-January 2007 a cyclone over the Netherlands generated a "European windstorm" with gusts of winds up to 202 km/h (129 mph) severing power to over 50,000 homes in the UK alone, grounding hundreds of commercial flights, slowing commercial rail traffic, closing several major highways across Europe, halting ferries on navigable water, and resulting in 47 deaths. Falling objects and motor vehicle collisions were major cause of the deaths. In many respects Hurricane/Winter Storm Sandy presents a very similar historic experience. There were also several airports being closed, a closed stock market (first time that had happened since 1888) initially with 8 million without power in 8 states, with 600,000 still without power 10 days after the storm hit. After the first few days of the storm the theater district was open in midtown Manhattan with $10 tickets being sold, since so many assumed Broadway and Off Broadway plays would be closed. Hurricane Sandy was typical in that it included many of the disconnects that too often surface quickly, and seemingly randomly. But FEMA and its supporting ESFs were able to mount an effective response, enhanced by the fact that there were no multiple events limiting FEMA's response.

As Sandy hit Connecticut, New York, and New Jersey it had mixed with other storms strengthening a bit, but it was the storm surge, working with the typical storm damage that affected housing the hardest. Many of the expensive but modest, two-story homes ($600,000 or more in value, at the time) were damaged by the storm surge and driving rain so that they were uninhabitable. Generally wet wallboard and trim has to be removed, and replaced along with furniture, appliances, and so on. The difficulty was that many had rented out one of their floors, so an uninhabitable house meant not only that the owners were homeless, but that the loss of expected rental income now put some of the homeowners at risk of foreclosure. It is true that the SBA does offer long-term low interest loans for just such circumstances, but the process takes months, not days. And FEMA payments to homeowners without enough insurance were capped in the low $30,000 rate, and these payments were far from assured, and as noted could be months away. Wind damage is usually covered by private home insurance. With issues like this surfacing, there was not much expectation that complicated shoreline mitigations issues could be well addressed. Fortunately, most homeowners appeared to have insurance, though flood damage and damage from storm surges are not usually covered by homeowner insurance, these needing to be covered by FEMA flood insurance. These highly technical and specific issues are mentioned here because they are the types of background issues that an Emergency Manager (or his or her advisers) must be aware of if their decision making is to be realistic and effective.

Emergency Managers must know a situation exists before they can implement response plans and allocate resources to a targeted and coordinated response. It is usually easier to detect that a winter storm has dropped precipitation on a given region, but its human impact is much more difficult to assess, especially when aerial over-flights and on-the-ground access to affected areas are limited regarding establishing an incident command structure. And even drones, where they have been deployed, have difficulty when the air is clouded with snow or rain. For example, the huge power outages that accompanied Sandy meant that power was out in many high-rise buildings, many of which contained elderly renters or condominium or cooperative owners who, lacking access to working elevators, were often not able to walk down 5 or 10 or even many more floors to refill prescriptions, replenish basic food needs, and related items. As previously discussed, in just a few weeks relatively healthy elderly or high functioning people with chronic conditions could turn into patients. Often, local paramedics had to enter these building and check on the elderly, their medical status and related items, but only after such at-risk buildings with at-risk residents were located, prioritized, and approached with appropriate transportation assets. Some of those buildings that had generators that powered elevators had no issues, as long as the generators worked and had sufficient fuel. Flooded streets that prevented people from going to work, stores, or medical facilities presented similar problems and had to be addressed as appropriate. Many affected areas were fortunate enough to have robust medical, EMS, and related systems, which were able to generate effective mutual aid responses in the early days of the response.[2] NDMS had DMAT assets available as backup support.

Tornadoes

Tornadoes are caused by intense thunderstorms and can have wind speeds that can reach 250–300 miles per hour or more. For a tornado to form, layers of air of different temperatures, moisture, density, and wind flow are required; tornados usually develop within intense thunder clouds (cumulonimbus) including unseasonable warm, humid air at the surface, with cold air at the middle atmospheric levels, and a strong upper-level jet streams.[141] Tornado season most often runs between March and August, but tornados appear any time of year; tornadoes tend to occur in the afternoon and evening, with more than 80% of all tornadoes striking between noon and midnight. Fortunately, of the 1000 or so that form every year in the United States just a few occur over land. The states with the greatest risk are Texas, Oklahoma, Arkansas, Missouri, and Kansa though in recent years tornadoes have struck in cities that are not regularly frequented by tornadoes, to include Miami, Nashville, and even Washington, DC (Fig. 12).

But it was the mile wide E5 (over 200 miles per hour) tornado that touched down in Joplin Missouri in 2011 killing 158, injuring over 1000 and wiping out a sizeable portion of the town that presents the most significant concern for Emergency Managers' decision making in the future (Figs. 13 and 14).

The one half to one mile or more ground sweep areas now being seen for some tornadoes is something unexpected and likely to change how tornadoes are defined in emergency management[2] and some have even tripled this ground width in recent years with one tornado covering approximately 2.6 miles in ground width. Tornadoes also appear accompanying some hurricanes. Hurricane Charlie, which affected Southwest Florida in 2004 was a serious Cat. 4 hurricane but did more extensive damage than initially predicted. Later analysis found that tornadoes were possibly responsible for much of these "unexpected" types of damage.

Early warning is key to survival in tornadoes, as citizens who have been warned can protect themselves by moving to structures designed to withstand tornado—force winds with building collapse and flying debris as the principal factors behind the deaths and injuries. Doppler

FIG. 12   Tornado damage can be sudden and dramatic.

FIG. 13   Large incident management operation.

FIG. 14   It's in the best interest of the emergency manager to help sustain local hospital operations.

radar and other meteorological tools have much improved the ability to detect tornadoes in the time available before a strike. Improved communications and growing public awareness have also been important to giving people better advanced warning, when a few minutes can separate death or injury from just having a harrowing experience (Ref. 2, pp. 46–9). Strengthened buildings (the previous discussions about building codes are once again in point) and safe rooms, which can be built into a new structure for as little as between $3000 and $5000, are the most cost-effective ways to mitigate the potential damage from a tornado. In fact, some of the same construction techniques that make buildings more wind resistant are those that are adopted in hurricane-prone areas; unfortunately tornado creation is often one of the more dangerous aspects of hurricanes. The Emergency Manager who has the time and staff resources to participate in successful community efforts to make these changes is fortunate. Again, the social and political skills of the Emergency Manager appear to be more important than his or her actual knowledge in these areas. Waugh and Tierney[67] cite an example of the benefits of timely tornado warnings that are both recognized as correct and to at least some extent followed:

> In May 2003, a strong (F3) tornado struck Moore, Oklahoma—near Oklahoma City in the heart of Tornado Alley. The National Weather Service issued a tornado warning fifteen minutes before the tornado hit, but Moore residents, who had experienced three tornados in the previous five years were prepared. The Emergency Manager activated the town's new warning sirens, and residents took shelter, according to the plan, in some 750 safe rooms, cellars, and other shelters. Thanks to an effective mitigation strategy, there were no deaths, and injuries were scattered. However, more than 300 buildings were destroyed, and at least 1200 more were damaged.[142]

A few decades ago, mentioning community organizing types of emergency management skills would no doubt have been met with derision, at least with some Emergency Managers. Fortunately, as a profession we have expanded both our reach and our talents, though budgets and access to wider levels of resources have not always followed.

Droughts and wildfires

Droughts are prolonged periods of water shortage compared to normal circumstances, are primarily due to prolonged periods of insufficient rain and other precipitation, or because of exceptionally high temperatures and low humidity causing agricultural difficulties and a loss of stored water. In very poor countries drought is associated with famine. In the US droughts can occasionally hamper commerce on major rivers, as the rare drought of 2012 halted barge traffic on the Mississippi, impacting billions of dollars of cargo and thousands of jobs. The severe and continuous drought in California between 2011 and 2018 damaged agricultural, required the Governor to issue water use restrictions and even kicked up giant clouds of dust that triggered asthma and contained arsenic. Most recently, by drying out normally moist tinder that used to act as a buffer for forest fires, dry tinder has been at least one factor contributing to the massive forest fire of 2019 in California, the worst ever experienced to date. Professor Edward P. Richards[232] recommends a concise summary on fire fundamentals.[144] Along with the other publications we have recommended "for the bookshelf" this eleven-page summary is one that Emergency Managers in wildland fire-prone areas should have or at least have read.

Internationally, many armed struggles have had their starst or have been seriously affected by water scarcity issues, such as in Syria[145] or Somalia and even the Israel and Palestinian conflict have water scarcity concerns. Certainly, we are not suggesting that the Emergency Manager become well versed in international diplomacy and history, though Col. Boyd of the OODA Loop would have found merit in becoming well versed in the areas. For Emergency Managers in rural areas or in areas prone to drought, it is important to stress and constantly keep in mind the power that water, or its absence can exert in many aspects of life and commerce. Of course, water is indispensable to human life and thought to be relatively plentiful in the US, it is limited and global demand for freshwater has been growing rapidly due to population growth, rapid urbanization and greater affluence. At the same time, accelerating global climate change and environmental degradation are altering the regional, seasonal availability and quality of water in the US as well as across the world. The resulting competition over water use may lead to conflict and sometimes violence, though researchers emphasize that it is rarely the lack of water as such that fuels conflict, but rather its political governance and management.[146] The National Integrated Drought Information System (NIDIS) has a comprehensive website https://www.drought.gov/drought/ that can be an excellent place to begin preparedness activities.[147] Fortunately, an Emergency Manager can also monitor the Climate Prediction Center of the National Weather Service providing reports on a weekly and monthly basis.[2]

Another often related aspect of droughts that the alert Emergency Manager must monitor are allied events that come with droughts, or that can be caused by them such as wildfires and dirt storms, which place time constraints on the Emergency Manager, as opposed to the more slowly moving, though frequently catastrophic drought events. But with global climate change accelerating hotter and drier as well as colder climate conditions, the sense that droughts won't happen in a specific area because they have not in the past may not be enough to give the Emergency Manager strong confidence. For example, extended drought conditions with commensurate failure to recharge in the Ogallala Aquifer area would be not just a state but a regional and national catastrophe affecting food availability, prices, and many related businesses and lives. Another drought with serious effects was the drought in Nevada in 2014 that exposed the lakebed of the Rye Patch Reservoir, which went as low as 3.5% capacity and caused the worst water shortage in the area in more than a century. At 174,000 square mile and running under South Dakota, Wyoming, Colorado, New Mexico, Nebraska, Wyoming, a Nebraska, Wyoming, Colorado, Kansas, Oklahoma, New Mexico, and northern Texas the Ogallala Acquirer is the largest in the world, supporting a huge portion of Americana agriculture.[148]

But as dangerous and continually worsening as draughts are in the US, in India up to 59,300 farmers have taken their lives since 1980 for reasons related to failing farms from draughts, though the numbers are no doubt considerably higher because of the shame associated with taking one's life.[149]

The "new normal" appears to be intensifying, widespread wildfires in the US as well as across the world. The National Interagency Fire Center in Boise, Idaho, correctly predicted for 2018 that warmer and drier-than-average temperatures, combined with large amounts of grass, below-average snowpack and increased potential for lightning was likely to create "above average to extreme" wildfire activity that year. Wildfires in August of 2018 have been validating that prediction with wild fires tearing across California, Colorado, New Mexico, and other Western states, chewing up bone-dry mountainsides, scorching buildings, and forcing hundreds of thousands of people to evacuate from their homes. Complicating the challenge for firefighters in several states is that over the last few decades, population growth and suburban expansion have led more and more people to build homes tucked into the very forests that are likely to burn. In California, three times as much land—120,000 acres—had burned by July of 2018, as compared with the same period in 2017, according to state data.[81] And now, some intense wildfires are starting to include "Firenados," whose frightening images are starting to appear all over the social media. A "firenado" also called a fire tornado, a fire devil, or a fire whirl can have swirling speeds up to 140 mph. They usually result from searing heat and swirling winds and can be miles high.

A final area of concern related to the growing wildfire menace is the unfortunate propensity of wildfires to contaminate water supplies, with the burned-out forests damaging the ability of the fire to percolate rain water so there is more runoff.[150] As Europe grappled with the near-record temperatures and continued droughts in Sweden had become the latest to confront a wave of wildfires as far north as the Arctic Circle, prompting authorities to evacuate some villages and to appeal to Norway and distant Italy for assistance in sending two planes to help dump water on the blazes as heat waves, drought, and blazes tormented Europe in July of 2018.[151] In the next year, Paris experienced temperatures in the 109 and 110°F range, a situation that was virtually unheard of until it happened. Intensifying worldwide wildfires and droughts point up the fact that despite having to extend his or her learning curve about rapidly growing natural hazards, the Emergency Managers still has to respond quickly as the "expected" events unfold regarding assisting with evacuations, working with hospital systems, the various federal agencies, the National Guard, the National Disaster Medical System, evacuation partners, and the mass of "routine" aspects of disaster response.

The fourth National Climate Assessment, Summary Sections can be accessed at https://nca2018.globalchange.gov/.[152] The massive report documents the expected growth of massive forest fires and their resulting damages, just as 2018 has included the worst forest fires in California ever recorded (this section is being completed in 2019, which gives every indication of being a record-breaking year, in itself). For example, in Peru the number of wildfires in the jungle is climbing at a rapid rate. Currently Peru lacks any type of wildfire mitigation effort or organized fire extinguishing resources. The impact these fires and deforestation have on the Amazon Basin, Andes mountain water supplies and global climate change is just beginning to be understood. The burden and the threat of wildfires place on California, western states as well as other Emergency Managers in the USand the world is impossible to quantify but to say that fire preparedness, response, recovery, and mitigation issues related to fires will no doubt expand at a faster rate than will emergency management staffing and resourcing.

A special difficulty that Emergency Managers face with wildfires is that each can be so different that it takes a completely different set of techniques to address. And fighting fires is best done through wise mitigation strategies before the increasingly hot and dry weather in many western states occurs. The most common call after a severe wildfire is to begin saying that certain areas should begin to restrict housing growth entirely, in fire-prone areas. But that is not likely to happen, so lesser strategies that include buffers of little burnable vegetation, roofs that are impervious to hot embers thrown out by approaching fires, and by residents making "defensible space" around their homes so that they can be more easily protected by firefighters. Of course controlled burns in some areas and other aspects of forestry management augment these more individual and local area strategies. The Emergency Managers in wildfire prone areas need a flexibility in decision making that accepts the growing risk of wildfires in fact their inevitability, and the need for locally tailored mitigation plans that will enable locally tailored response plans when wildfires occur.[153]

Heat waves

Heat wave (or extreme heat events in "CDC speak) conditions are defined by summertime weather that is substantially hotter for a location during a specific time band. Heat wave events can rise suddenly, putting the Emergency Manager in the position of having to react quickly and accurately to limit morbidity and mortality; complicating the response is the fact that local, state, and federal health and medical agencies generally have lead for these activities. Even though heightened awareness of global climate change has brought heat waves "into the news" across the US and the world, when temperatures rise to higher than normal levels during summertime, myths and half-truths persist. These can have damaging effects, especially when the reporting is from national sources, and local areas just cannot be adequately addressed in general reporting. One frequently misunderstood point is that people do not tend to die just because temperatures are high, even much higher than normal, for a given area. For example, people do not die at higher rates in places that normally have the highest absolute temperatures, places like Arizona, for example. In fact, in places like Phoenix, where temperatures can frequently exceed 110–115°F or more in summertime, almost no one dies of heat-related causes. The mass media as well as the general population are generally well attuned to heat-related issues so that air conditioning and checking in on the elderly or those physically impaired is a normal aspect of living. People's bodies have also acclimated to the heat in ways that they may not have to in other areas. Another area of confusion is between absolute temperatures and heat indexes, but in areas more attuned to heat events, there is little confusion.

Heat indexes are composed of absolute temperatures and humidity levels. For example, an absolute temperature of 118°F in Nogales, Arizona is likely to be accompanied by lower humidity levels and will not be deadly as a 105°F in Chicago, which is virtually always accompanied by substantial humidity. Over 700 died in Chicago's 1995 Heat Wave, peaking after the second day of the event, the deadliest US heat event in decades, but there have been others. In San Francisco, in almost 150 years of record keeping, it has never been as hot as it was on a September day in 2017. Amid a brutal heat wave that had affected California for a week while intermittent knocking out power to thousands and fueling more than a dozen wildfires, downtown San Francisco hit 106°F degrees, not normally a city accustomed to such high heat levels.[154] But the event was not deadly as it occurred at the end of summer, when people's bodies were already acclimated to the heat, and was part of a continuing heat wave, with no quick and unexpected run up in temperatures of the kind that made Chicago's 1995 heat wave so deadly.

The deadly potential of heat waves cannot be minimized by the Emergency Manager, no matter where he or she works. Over the last 30 years heat waves have killed more Americans than any other weather-related hazard.

In 2003 approximately 70,000 died in deadly European heat waves, many of which saw relatively affluent elderly dying when their children and grandchildren were on vacation, away from the area. Air conditioning could have been helpful, but has not been prevalent in Europe, though that situation is rapidly changing as the weather in Europe grows more extreme. The Europeans have learned to avoid the huge death rates suffered in 2003, even when facing temperatures as high as 110°F in Paris in 2019. Air conditioning, once mainly a luxury in hotels is now becoming more commonplace, as well as are much higher levels of vigilance for the elderly and others more vulnerable to the ravages of extreme heat. And unfortunately, responding effectively to a rapidly forming and deadly heat waves requires more in-depth knowledge of surrounding events, than just having an effective preparedness plan based on the best in prevailing national heat wave plans. Fortunately, there are excellent resources, discussed later, for making the Emergency Managers a modest "expert" in heat waves preparedness and response. Much easier to make quick and effective heat wave decisions when the issues that are likely to appear, and their most common causes are already in the Emergency Manager's mind, checklists and plans; and the relationship with health and medical agencies is well established by mutual trainings and exercising.

Urban heat islands

The long-established concept of the 'urban heat island" has existed for over 100 years in the US, is pervasive in the European literature on heat waves, and applies, to a lesser extent, to urban areas in the developing world. Higher temperatures can affect communities by increasing peak energy demand, air conditioning, air pollution, and heat-related illnesses and mortality (Fig. 4).

The urban areas tend to be warmer because of their "masses of stone, brick, concrete, asphalt and cement." These darker surfaces, which can absorb more solar heat during the day, radiate that energy back into the urban environment at night. There also tends to be fewer cooling trees, bushes, and vegetation, including diminished wind and cooling air currents in the urban heat island parts of cities. Although heat island, inner-city neighborhoods can be populated by affluent residents, the bulk of attributes of inner-city residents categorize the poor. The indicators mostly include poverty, social isolation, social class, race/minority status, crime, poorer housing, inadequate health care, high divorce rates and broken families, mobility limitations, and the prevalence of Single Occupancy Residences (SRO) which typically have the highest rates of social problems and the highest mortality rates during heat waves. It should be noted that urban heat islands are not just a product of larger, primarily older, cement and asphalt hardened urban areas, they can include smaller, but just as deadly neighborhoods in smaller cities, towns, and even inner-ring suburbs whose demographics most often mirror those of the larger, inner cities areas (e.g., East Liverpool, Ohio). The urban myth that the medical examiners often entered units in which an elderly resident (or two) had died during the heat waves, but with the air conditioner(s) turned off are sometimes true.

Wise Emergency Managers have copies of some of the best and latest preparedness plans available as teaching tools as well as potential templates for their own plans and updates. In the heat wave area, the heat wave plans developed by Philadelphia, Toronto, and Chicago are good places to start. *The Excessive Heat Events Guidebook*, 2006, pp. 22 (available electronically at http://www.epa.gov/heatisland/about/heatresponseprograms.html)[155] lists useful inclusions for an effective planning preparedness process (based to a great degree on the Philadelphia and Toronto heat wave plans):

- Access to weather forecasts capable of predicting 1–5 days in advance of heat waves
- Coordinate transfer of weather forecasts by heat wave program staff
- Quantify estimates of heat wave health impacts, heat-related deaths, info on high-risk individuals and an accessible record on facilities/ locations with concentrations of high-risk individuals.
- Coordinate public broadcasts about the timing, severity, and durations of the heat wave
- Operate informational phone lines
- Designate public cooling centers, extend operating hours of air conditioned centers
- Add extra staff for emergency support services and community liaisons with churches/synagogues/mosques and social organizations
- Directly contact and evaluate environmental conditions and health status of known high-risk individuals and locations like to have concentrations of these individuals
- Increase outreach to homeless and establish provisions for protective removal to cooling shelters
- Suspend utility shutoffs
- Reschedule public events as needed
- Develop and promote actions to reduce the effects of urban heat islands

Individuals at highest risk of becoming ill or dying during heat waves are the very young, the elderly (socially isolated, and sometimes without access to air conditioning), bedridden, and with ischemic heart disease or other chronic conditions, the poor, minorities, and those taking certain medications such as neuroleptics or antiparkinsonian agents. Often a constellation of these variables appears in those residing in Single Room Occupancy units in urban heat island neighborhoods as described before.[156]

Viewing Fig. 4 above, it is clear that the existence of large amounts of elderly (75+) residents in an area is a strong factor in heat-related mortality and morbidity. But hidden in that risk factor is the physiological tendency for the elderly to lose some of their sense of thirst and body temperature regulation. In practical terms this may mean not turning on air conditioners (especially if poverty is a factor) even though inside temperatures rise to dangerous levels, not feeling the need to adequately hydrate and beginning to experience heat-related illnesses. The fact that many thousands of frail, middle-class French elderly died in an unexpectedly severe heat wave may have been because their lack of hydrations caused them to be "dehydrated…. with evidence of kidney failure."[157] India, with its masses of poor, with recent 111°F days by lunchtime and as high as 118°F. has seen its mortality rate in these areas rise to as much as 43% higher than in more temperate days.[158] And Europe, not normally an environment that has fostered heat waves in 2015 experienced record shattering heat in German, France, and the Netherlands.[231] This condition repeats itself in virtually all serious heat waves that affect the elderly. Recent research shows the elderly response to heat includes significant impairments to judgment, making issues of hydration, air conditioner use, and related concern even deadlier. Additional research shows even the young and strong, when working and playing outside in extreme heat have a surprisingly lower threshold for heat-related damage than previously believed. Emergency Managers need to assess the demographic and urban heat island-type areas as part of their normal preparedness plans, to prejudge heat wave risk, well before the rare but all too frequent heat wave conditions appear. As a final point, it is too easy to assume that ESF#8 will have worked with hospitals, mutual aid, and even NDMS DMAT assistance if necessary, public media, local EMS, and fire departments to trigger the response immediately. The authors have seen quick response, even with the existence of high-quality heat wave plans hindered by powerful medical personalities that have not participated in preparedness for heat waves, yet begin to "customize" the response. This is where the Emergency Manager's communication and decision making skills will come to the fore, with any luck, that is. Though heat wave exercises are spending scarce time on a rare event, the deadly nature of the events and the increasing chances of more events because of global climate change do indicate that occasional heat wave exercises are realistic. And besides, a well-drawn heat wave response exercise is also a medical emergency response exercise, by another name.

Technological hazards

Unfortunately, as FEMA wisely addressed in 1993, it is impossible to conduct research on many technological disasters before they happen.

These events need to be studied postdisaster, and researchers will always be behind the curve as technology advances.[159] The Emergency Manager must know his or her area, work with private industry, and be quick to realize that natural disasters can also have technological aspects that can be stronger in impact than the original disaster[2] define technological hazards as inevitable products of technological innovations and human development to include structural fires, dam failures, hazardous material incidents, and nuclear and radiation accidents. They result from the failures, misuse, or unintended consequences of engineered structures, technologies, manufacturing processes, or other aspects of modern life. They observe that these tend to be less understood than their natural counterparts and also are trending in a direction of increasing frequency as the scope of and dependence on technology expands with the transportation sector being the most likely sector to experience serious events. Waugh and Tierney[67] stress the inherent difficulties in responding to technological events:

> Because of the complex underlying human factors, it is more difficult to assess the level of risk from man-made hazards than hazards from natural causes. Nevertheless, mitigation planners can develop lists of potential man-made hazards in their jurisdictions, and they can review public and private emergency and contingency plans and reports, including radiological emergency plans. Radiological plans are a good example as they are FEMA funded and exercised, with updated annual copies sent to key ESFs, developed by state and local emergency agencies with the cooperation of the local nuclear facilities. Other examples are the emergency plans of facilities that use hazardous materials, emergency plans for chemical plants, stockpiles, toxic release inventory reports, and statewide domestic preparedness strategies.

Much of what an Emergency Manager needs to do in responding to most technological disasters is to recognize that he/she may always be behind the understanding curve regarding hazards and risks. Having knowledgeable and available sources of information in the various technological areas would be ideal, but when the sources are most needed, in a response situation, tends to be the time when they are hardest to secure. Recognizing this, Emergency Manager need succinct, quickly retrievable, though almost by definition out-of-date plans preparedness information. This is more of a problem for local Emergency Managers with both short staffs and short funding, and a bit less so for federal, particularly FEMA, Emergency Managers with a wealth of well-informed and accessible SMEs. But just because the risk of most technological hazards is remote does not imply they can be virtually disregarded. For example, as discussed previously, in rural areas the largest employer, and also often the chief source of technological hazard is the local power utility, be it coal fired, natural gas, nuclear, or in transition. Rural highways especially state and federal trunks, or railways are obvious sources of potentially serious hazardous materials risk. And, by their isolated, rural location some communities have the economic good fortune to have plants or industries posing a hazardous material risk, and the additional resources to address potential risk, working in close cooperation with the plant or industry. When viewed in an economic perspective, most (though clearly, not all) rural communities would gladly accept the risk of hazardous material incidents for the economic benefits that accompany the risk.

A quick review of "Lessons Learned and Not Learned" from Three Mile Island can further frame decision making quandary that technological hazards can present Emergency Managers. Most of these findings related to preparedness but serve as valuable perspective for Emergency Managers at all levels[160]:

Lessons for industry based on the Three Mile Island (TMI) reactor disaster

- Responsibility for operating belongs to managers, not to any federal or state regulations. Regulators try to establish an envelope for safe operations, but it is management's responsibility to reduce the envelope if they see fit. At the time of the accident, roles and responsibilities had not been adequate determined.
- Adequate risk assessment had not been conducted at Three Mile Island.
- Hazardous technologies require highly competent managers who have confidence in their continued respect and employment. Simple and direct lines of control need to be established but were not.
- The competence of the operating crew is critical, but the utility did not pay for competent managers.
- The growing influence of lawyers and the courts are no substitute for adequate investment in effective regulations, staff, and required safety infrastructure. It is too easy to use the law as an excuse for doing little else.

Lessons for the regulator based on the Three Mile island (TMI) reactor disaster

- Accidents must be prepared for, trained for, and planned for. At TMI they were not.
- Problems need to be shared by the various levels of regulators. Of course, industry should be pushed to accept more responsibility, but regulators must share a philosophy to be effective.
- The courts will have growing influence, so regulatory actions need to be clear and defensible to the extent practicable.
- The public must be educated about risk.

Terrorism

After the World Trade Center attacks in 1993 and the 2001 with all of the emphasis on terrorism (as discussed previously) the lives of Emergency Managers have been complicated both by the additional mission (which they do not control) and in the general tightness of funding and lesser emphasis on traditional, naturally occurring or technological disasters. Disaster managers with large or complex urban areas under their coverage or rural areas with select, terrorist targets will need to emphasize terrorist concerns, though most often rural Emergency Managers can usually continue to address "all of the other" standard disasters. But serious terrorist events, which generally fall under the FBI's responsibility, and emphasize strong security aspects, usually still include the heavy involvement of many standard emergency management roles and assets.

In a chemical event, standard hazardous materials skills will be useful, but will not necessarily have been designed to detect chemical agents most likely used by terrorists. Agent identification will remain difficult, especially if the agents have been delivered by covert means. Biological agents can be identified by existing public health monitoring systems or by local, state, or national networks working with epidemiological tools.[2] Regardless, owning and having staff trained in the use of expensive Polymerase Chain Reaction (PCR) test kits remains an important "backup" to local Emergency Managers. It is clear knowledge to respond to the terrorists' use of Weapons of Mass Destruction will come from previous US and world events, various types of scientific research and expert materials, exercises, and related training. But responding to a significant terrorist event will only be done in conjunction with a coordinated joint federal, state, and local approach, under the authority of the FBI and whichever elements it delegates. If the event is large enough the US Department of Health and Human Service can send elements of the National Disaster Medical System to augment local public health and medical response assets along with elements of the US Public Health Service Commissioned Corps (Fig. 15).

The irony is that despite the strong emphasis on terrorism, local, state, and federal Emergency Managers will still use the same mutual aid EMS/HazMat, hospital system, transportation, environmental assets, and so on, response and supports assets under the standard all-hazards approach in dealing with evacuations, injuries, contaminated scenes, communications issue, transportation issues, and public information strategies that accompany larger terrorist events. Since identification of chemical and biological agents is basic to responding to these events, the wise Emergency Manager cultivates strong local and state public health and environmental relationships. However, agent identification and the involvement of local public health and environmental staff and even staff from the Centers of Disease Control and Prevention will proceed with or without the involvement of

FIG. 15   How will the emergency manager handle this type of challenge?

the local, state, or federal Emergency Managers. Staying in the information loop with these entities is useful to the extent that state, local, or even
federal emergency management can be of great assistance to external staff,
such as CDC, regarding securing them accommodations, local transportation, situational briefing, phone/email lists, and so on. This is the time
when constantly updated local information and staff access lists prove
their worth.

The vast scope of terrorist incidents precludes more specific instances
of information and processes that may help to improve emergency management decision making. However, despite the potential overemphasis
on terrorism that many smaller and middle-sized emergency management
agencies may face, it is still useful for the Emergency Manager to produce
his or her own models of terrorism response, as Col. Boyd may have observed. To this extent the *ABS* (*American Behavioral Scientist*) special issue
on terrorism though nearly 20 years old, is still quite helpful, and a compact, quick read.[161] The next book is quite different and present a detailed
discussion of terrorism, some of the groups of note and much case study
information (Bolz et al.'s *The Counterterrorism Handbook: Tactics, Procedure,
and Techniques* [162]; The final book recommended is Christopher Dickey's

Securing the City: Inside America's Best Counter Force—the NYPD.[163] Dickey is a journalist, and not a terrorism expert per se, but his descriptions of New York's counterterrorism structures and efforts are very helpful regarding every terrorist's favorite target. Also Richard Priest's *Hot Zone: A Terrifying True Story*[164] is a riveting story of how a deadly virus from central that kills 90% of its victims, and for which there is no treatment or immunization appears in the suburbs of Washington, D.C. This little paperback was made required reading by Frank Young, M.D., PhD, former Director of the Office of Emergency Preparedness, the original little agency from which the Office of the Assistant Secretary for Preparedness and Response was derived.

Common mistakes in decision making during events

This chapter will combine some of the experiences the authors have had with poor decision making during disasters with a selection of mistaken decision making from after-action reports as well as the disaster literature.

Confusing success in procedural/bureaucratic processes with substantive success in the mission(s)

It has always been much too easy to feel as if you are accomplishing something, staying ahead of the disaster curve, when you are spending lots of time scurrying from meeting to meeting, doing summaries, situation reports, and meeting all the rest of the bureaucratic requirements emergency managers must live with, regardless of level or agency type. And this feeling that "You are doing something of value" can get much easier during a disaster that is very dangerous in scope and covers a larger geographic area. But the Emergency Manager must constantly remind himself or herself that we must measure ourselves against the only two objectives that we should be concerned with: (1) gathering appropriate and sufficient data to determine which missions are required to be done, and under which priority and (2) implementing these substantive missions affecting residents, infrastructure, environmental, and local economic/job-related needs. Properly staffing EOCs, preparing and submitting timely situation reports, meeting notes, good relations with the Federal Coordinating Officer, giving your own correct and timely reports, meeting the ever-increasing demand from "higher in the food chain" for more and more data, and all the rest of the nonsubstantive requirements are to be viewed as supporting a bureaucratic structure that will help us implement the substantive missions. But these are only tools and procedures, not end products (at least from the viewpoint of the Emergency Manager) or the victims of the disaster or emergency.

As a final point, the emphasis on organizational and procedural change that accompanies most political change, or the sad piece of wisdom that holds "If you don't know what to do, or what you are doing, then reorganize," is still true and will probably always be. Don Kettl suggests that under James Lee Witt, a skilled "Southern Emergency Manager" appointed FEMA Director under President Clinton, "success in managing FEMA flowed from the leader's ability to lead...Restructuring cannot substitute for leadership...Structure matters. But leadership counts for more." (These comments appeared in a 2009 Department of Homeland Security, Office of the Inspector General Report: "FEMA: In or Out", appearing in Ref. 2.) (Fig. 1).

FIG. 1   Politics and emergency management is a delicate dance.

A brief case study of failed decisions making by a FEMA manager confusing bureaucratic processes with "substantive success in the missions(s)"

Once again, using Hurricane Katrina as a backdrop to typify the points we are making, we focus on the City of New Orleans: during the second and third day of the disaster, when water was still flowing over and under the breached dikes in the New Orleans's Ninth Ward, those that could evacuate were leaving the City or relocating to shelters (The Holidome and the Morial Convention Center, among others) but thousands were still stranded on their roofs of their homes or apartments or in them and many other thousands were crammed in sometimes dirty and dangerous conditions in the various shelter locations. And recall, adequate security for all of this, an overburdened airport evacuation, almost nonexistent buses, helicopters, and so many more necessary response tools, supplies, and materials were either in short supply or did not exist. And of course, the devastation spread across 6 states at various levels of destruction, further straining normal local and regional mutual aid response resources.

In the middle of all of this widespread devastation, a member Emergency Support Function #8 (ESF #8), public health and medical systems, received a call from a niece attending a local university. She had just been called by a nurse friend who was stranded on a New Orleans area hospital roof with 12 or 13 brittle patients, most of which were on portable IV lines and most if not all of which were in danger of dying if they did not receive their required medicines, oxygen, and various other treatments. By this time the nurse, whose cell phone was rapidly draining of power, who was in charge of them was extremely fearful and angry, to such an extent that she felt that the totality of local, state, and federal emergency response agencies had not only failed her and her patients, but the city and state. This situation of both healthy individuals and less healthy patients who have their access to required medicines and treatments damaged because of a disaster, and are now at risk of death is a commonly recurring one in virtually all extensive disasters. Because of flooded roadways and a lack of appropriate evacuation assets the Lead ESF #8 staffer approached the FEMA Night Operations Chief and made him aware of the dire circumstances now facing the nurse and her patients. But first the nurse was called (her cell phone number was supplied by the staff member's niece) and was told that the situation was being worked. Her anger and fear were at such high levels that it took some time to let her vent and provide her location, some briefing specifics about her patients and their declining physical status. Kahneman has frequently observed that it is much easier to analyze decision making in others as opposed to deeply understanding our own decision making processes.

The FEMA night Operations Chief was hurrying to what he felt was an important meeting and said that addressing the patient evacuation issue needed to wait an hour or two, to be addressed at the next all-ESF briefing. He observed that he did not have staff available to get on the issue immediately, again reiterating that it would have to wait until the upcoming meeting. The ESF#8 staffer quickly looked for a landline, hoping for better reception than on what he was experiencing on his cell phone. The FCO was not on site, so "going over the head" of the night Ops Chief would have taken too much time, and would have posed its own difficulties. Fortunately, the adjacent Parish had a rotary wing available and sent it out to the hospital roof, immediately after the conversation. As noted, the nurse in charge had already been contacted and was packaging the patients for air evacuation, as quickly as she could. The Lead ESF #8 responder never had a chance to track the patients, but not hearing again from the nurse, knew that the situation was resolved with no fatalities. The FEMA night Operations Chief was informed of what has just transpired, but apparently had forgotten about the issue, as he rushed off to another meeting.

The FEMA night Operations Chief was not an incompetent employee, in fact his reputation for competence landed him in this key position. The difficulty was that he was so caught up in the hale of meetings and briefings and reports that he was not able to focus on a true life and death situation presented by 12 or 13 brittle patients stranded on a roof. The FEMA staffer who could have advised this harried night Operations Chief to respond immediately did not appear, though such an individual may have been at the upcoming meeting. Stallard and Sanger[6] wrote of the value of wizened staff members advising top level leaders of impending bad decision making. Unfortunately, the availability of such skilled staff just when they are most needed is most often a matter of chance, and not necessarily of truly wise institutional planning. The Military Decision Making Process does recognize the value of this restraining skill and of its existence in the normal chains of command. Unfortunately, in this case good advice was given, but not followed. It is too easy to forget that even the intense mix of talent and dedication demonstrated by so many Emergency Managers on a day by day and night by night manner is sometimes not enough to assure that key decisions will be made correctly. Emergency management is not a career field that draws those most interested in money. We should never forget that "most of us are all about the mission, even when we are screwing up."

The problem of making good decisions in a group

Group decision making is at the heart of emergency management decision making though we all recognize, or at least should, that it is quite a bit more complex getting good decisions out of a group than out of an individual or two. But the increased complexity of dealing with more than one individual in the decision making process can be more than balanced by the larger amounts of additional information and different perspectives supplied, by varying, and even clashing viewpoints and perspectives of people with different talents and personalities, at different levels of the chain of command. If it seems like we are beginning to expose the reader to the huge complexities as well as the huge benefits of something we all take for granted, group decision making that is the correct take away. The point of this section is not so much to learn about situations in which bad decision making is likely, but rather to consider the various aspects of the group in any decision making situation. This means that Emergency Managers need to keep uppermost in their minds, and many do understand that personality clashes, people whose confidence in their emergency management decision making is not warranted, and a variety of individual and group circumstances that can enhance, minimize, or even destroy the value of the decision making group we are dealing with, need to be considered as decisions are made or postponed. We'll include a number of sections dealing with different aspects of group decision making. Again, the field is vast, and we will not try to summarize it here, but rather to pull out a few examples of what we have experienced or learned. Some of the examples of the types of bad decision making we have experienced will include some of the key decision making concepts that were addressed in Chapter 1. The key concepts bear repeating in this problem-oriented context.

Group decision making

The point of this discussion of group decision making will not necessarily be to review the voluminous thought and research conducted on group decision making, and certainly not to delve deeply into the key findings of group psychology, though that would be useful reading for any Emergency Manager, or for any person working with others, for that matter. Rather, we hope to suggest that the Emergency Manager consider the decision making group as something that can seriously hurt or help in the process of making good, well-informed decisions in a timely manner. This implies that the emergency manner at least recognizes the potential values as well as the potential pitfalls of the groups they select, direct, or participate in, and to take nothing for granted. And recall, we have all witnessed an otherwise effective briefing destroyed by the animosity between two, otherwise talented and seemingly wise participants who just happened to dislike each other, or a member who just can't keep from talking, and talking. To a great extent we will present issues in group decision making that we have seen appear and reappear, under many different circumstances in emergency situations in the emergency management literature as well as some classic findings, as exemplified by "Group Think" as follows.

Group think

The term "Groupthink"[37] has now joined the vocabulary of emergency management decision making, as it has in so many other areas of organizational and institutional decision making. And for the most part, this is a good thing. But, as noted in Chapter 1, Group Think does not necessarily imply that decisions made in groups are necessarily suspect, or more prone to error than individual decision making. Janis found that members of high-status groups often tended to mute their criticism of what is being discussed, regardless of their views, because they prize so highly their membership and the respect of others in the group. Of course, that kind of thinking got our nation into the Vietnam War, the initial invasion of Iraq, and many other fruitless ventures. In a practical sense, this means that the Emergency Manager must guard against groups of highly talented old friends and colleagues that all seem "to think alike on the important issues." That could be a good thing, but likely the close relationships and the strong respect for each other may stifle some of the creativity and aggression that the effective Emergency Manager should display in some instances. There is a time to challenge a dangerous or ineffective course of action, no matter who proposes it, nor who else supports it. In these instances, silence is not an acceptable course of action, it is just an example of bad decision making.

The problem of professional judgment in decision making during events

The experienced Emergency Manager knows that the advice of even the most trusted Subject Matter Experts (SME) can give assessments or estimates that are widely divergent from those of their colleagues or even from their own past assessments and estimates in similar situations. In fact, most everything that has been presented here demonstrates that professional judgment as well as virtually every other type of judgment should be examined if not be outright suspected of being biased or at minimum inaccurate, as a starting point. In emergency management this fact imposes certain costs and constraints that can, on occasion, be simply dangerous if overlooked or ignored. For example, the Lead, Regional Administrator for HHS can estimate, on the first day of an event resulting from a Category 3 hurricane, that the National Disaster Medical System needs to send (or to have predeployed) up to 200 National Disaster

Medical System staff to Puerto Rico to conduct medical system assessments and to provide primary care for those unable to access their existing practitioners. Of course, this does not include the potential of upper management override, for a variety of reasons, occasionally. While this same person may have estimated over 100 to be predeployed to Puerto Rico on a roughly similar situation a few years later.

These same inconsistencies can appear in the size of debris removal contracts in various events, the amount of water or blue FEMA tarps, or just about anything else. And this is to be expected in complex decision making or estimates. But it needs to be stressed that professional judgment that is less than precise is not necessarily a deadly flaw, just something an Emergency Manager must track. What is certain is that debris contracts, tarps, water, MREs, chain saws, various sizes of generators, buses, National Guard and Department of Defense support elements, NDMS DMATs, and many other required items appear like clockwork, as does the opening of a Joint Operations Center and so much else. It is just the timing of the asset deployments, the numbers, the supplementary staffing, and the locations that are subject to normally fallible judgments in decision making. Recall the earlier findings of Kahneman et al.[26] regarding the frequent and general unreliability of so many decisions made by professionals in their fields of expertise. A copy of their article "Noise: How to Overcome the High, Hidden Costs of Inconsistent Decision Making" should probably be on the book shelf of every Emergency Manager and reviewed every so often.

Potential problems in dealing with FEMA's Federal Coordinating Officers

Early in the course of any declared disaster, FEMA will appoint at least one Federal Coordinating Officer (FCO). These field managers have usually been men (though this has been changing) most often, at least in their 50s, who have had strong or even distinguished military, fire service (often as paramedics), enforcement, or even federal careers as upper level managers. These individuals are usually of the "can do" and early action variety, and are, by definition not always as well suited to the longer, grinding aspects of recovery with its messier issues of shelters, dealing with powerful nonprofits such as the Red Cross or the Southern Baptist Food Services, that are anything but quick and easily settled issues and processes. Few FEMA staff members apply for and are promoted to these GS 15 grades. Many FCOs have had justifiable confidence in their leadership experience, but sometimes little experience or relevant knowledge in the substantive areas for which they will have control during the disaster response and into the recovery periods, for the disasters they are assigned. So, they've been "around the block," just not necessarily the block they are now working on. In fairness, FEMA training includes substance and process materials on each of the ESFs, updated frequently. We've discussed that leadership, especially newly appointed leadership can easily fall prey to overconfidence, a situation easily complicated with a lack of specific experience with many substantive issues such as environmental, public health and medical systems, US Army Corps-related, and similar substantive issues. To this effect, as experienced leaders FCOs will probably be more interested in avoiding huge mistakes than in holding out for the highest quality or response. This does not imply FCOs want sloppy response activities, just that they are more attuned to avoiding massive failures as are many other federal and even state employees. But in fairness, these individuals are usually quick studies and can be expected to recall what they learned in their training courses and what they have experienced in other disasters. Nonetheless, their expected confidence in most areas can be a benefit as well as a drawback. Also, FCOs are particularly focused on Regional and Headquarters recommendations so as not to prejudice their chances at future deployments. This is not a cynical comment, as FEMA needs effective and cooperative FCOs for a variety of reasons, all of them obvious.

Federal Emergency Managers reporting to FCOs, because the FCO is technically in charge of managing, calling meetings, requesting staffing, and funding the FEMA disaster response, need to travel a narrow path. On one hand, the early days of a disaster can more easily get FEMA approval of emergency expenditures in health and medical, environmental, and related items. As the disaster progresses, federal Emergency Managers from most of the ESFs supporting FEMA might benefit if their reports are short, pointed, and honestly indicate good progress (something all benefit by). Looser reports may draw FCO interest, not always a wise option. On the local and state levels, the situations are different, where the intent will be to draw FEMA interest and funding. And this is all is complicated by the growing amount of automated and highly detailed data that is now generated from so many sources. This data is prized by those highest in the chains of command, and their political leadership, but can become such an imperative that sometimes the direction and intensity of the response can be hindered by the upper level pressures for more, faster and more detailed data. This is a constant theme. Of course, the FCO has to relate to the upper levels of FEMA and that implies political exposure, always a difficult situation. Federal, state, and local Emergency Managers will have varying levels of political exposure, but all FCOs will have extensive political exposure in almost anything they do or decide not to do. To work effectively with an FCO these facts must be uppermost in the Emergency Manager's mind.

Working with the FCO, as noted, can be a dual edged sword. For example, early in a disaster hiring now unemployed locals to do environmental cleanup can meet the goals of helping to jump start a local economy (assuming local hiring), while also reducing environmental issues, clearing roads for better transportations, with a host of related benefits. Getting an FCO to make this recommendation "up the chain" can be very effective. However, if the FCO is unable to get approval for these types of expenditure, the federal Emergency Managers working with the FCO now have a newly educated, newly enervated advocate who will no doubt begin requesting more information and more involvement, serve his or her organization's needs, but not necessarily those substantive (or organizational) needs being advocated by other federal Emergency Managers supporting the FCO. Of course, this is an extreme example, but the point is that failed requests for even slightly "out of the ordinary" requests for mission assignments always risk enhanced interest from many who would tend to complicate the lives of federal ESF Emergency Managers. On the other hand, good communication with the FCO and convincing him or her of the value of the mission to the victims, the state and local areas are always a good strategy, if it is successful.

In this section everything has been primarily written from the perspective of federal Emergency Managers leading or working in one of the federal support agencies performing ESF functions. Even a brief review of Chapter 1 should convince the Emergency Manager at any level to try and put him or herself in the place of the FCO and the complex task that they have, if he or she attempts to do it well, as most no doubt will. These complex, sometimes even contradictory, tasks will almost always include recognizing the FCO's constant need for more and higher levels of data and information to feed his FEMA Washington Office, managing the special needs of his FEMA Regional Office (never forget that disasters are primarily regional events from FEMA's perspective) along with the FEMA Regional Director, various states and cities that are central to the response. Public Relations staff seeking accurate information of the type that puts FEMA in the best possible light, as one would expect is also an ongoing priority. Some empathy to the hard life the FCO experiences can make life easier for him or her and the federal Emergency Managers that deal with him or her.

Failing to consider predeployment of staff and assets when appropriate and practical

In most disasters, certainly in larger catastrophic ones such as massive hurricanes, forest fires, floods, and winter storms, the response starts well before state, local, and federal EOCs begin opening. Depending on the type of disaster many pre-event activities such as logistics, administrative, and even financial asset deployments start well before the disaster or the NSSE begins, assuming an event that all can see developing. These comments mainly apply to federal Emergency Managers in the ESFs, who usually have more substantive travel resources and flexibility to consider predeployment activities than local or state Emergency Managers. For example, larger hotels and high schools, and even some airport facilities in at-risk, hurricane areas have been built to 140–50 miles per hour standards. Predeploying FEMA communications/IT staff and other federal administrative staff to "ride out the storm" to get a start on either developing an initial operations center in hotel meeting rooms can produce early information on an area's past event status, including coordination with local emergency management liaisons. More recently, the use of drones will both complicate and assist predeployments.

With all of the normal and expected pre-event preparedness it is very easy for the busy Emergency Manager to fail to consider additional items, staffing, services, or equipment that can be predeployed. This in no way minimizes FEMA's often considerable pre-event activities. For example, 2800 activated National Guard troops, 13 open shelters, tons of water and pharmaceuticals, chainsaws among other equipment and activities were made in preparation for Hurricane Florence in 2018, in North Carolina; or the estimated $2.4 Billion FEMA spent in 2017 for Hurricanes Harvey, Irma, and Maria, already being accomplished routinely, including the existence of "advance contracts" for expenditures for products and services such as prefabricated buildings, the "blue tarps," and food and inspection services. And, when the process is working smoothly and is adequately funded (sadly, not always the case), FEMA works with states and local areas to integrate materials from after-actions reports to more smoothly facilitate mutual aid, distribution points, and related adaptations, though such preplanning and preparedness was not much in evidence regarding Hurricane Maria in Puerto Rico in 2017. The fact that Puerto Rico did not request mutual aid support before Maria's landing should have alerted FEMA to question why this was not being done, and begin some of those initial support activities itself, at least in life safety, communications, transportation, and power related areas.

These predeployed staff could have assisted state/island staff in securing mutual aid staffing and assets with less time passing and could have helped FEMA Incident Management Teams to begin gathering (and sharing) critical status information. Given that emergency medical treatment, especially during the first "golden 24 hour" period after an event, can really benefit by having a number of advance elements of the National Disaster Medical System, disaster medical assistance team (DMAT) ready to treat patients, conduct medical needs assessments, or augment local treatment capacities. This assumes a selection of safe locations are available for public health and medical staff to ride out the storm that are safe, yet accessible to key areas of need after the storm has passed. Once this has been done, the decisions to predeploy can be made with more confidence.

Many kinds of staff and assets, but particularly NDMS DMATs, NDMS Medical Assessment Teams, occasionally FEMA Incident Management Assistance Teams, rotary "fly-over" aircraft, and many other predeployed assets should be considered by Emergency Managers at all levels. Time lost at the front end of an event cannot be made up, medical treatment not rendered and the potentially deadly problems that can result cannot be easily fixed later; assessments not done early will never be as effective. There are other types of staff and assets that can be predeployed that can really help the timing and effectiveness of a response. For example, FEMA in April of 2018, in preparation for the upcoming hurricane season created a number of FEMA Integration Teams, which were permanently deployed to work with state officials and provide technical and training assistance, according to the North Carolina Emergency Management Agency, the first state to benefit by this type of predeployment. Also, there are the potentially tens of thousands of workers that can be predeployed in impending hurricane events by the energy companies. In preparations for Hurricane Florence by Duke Energy, which covers some four million customers in the Carolina, some 20,000 workers were placed in hotels, staging areas, and—for those employed local—in their homes to be immediately called upon once it is safe to do so. An additional 20,000 were dispatched to other utilities in the area. Interestingly, in Puerto Rico in 2017 the Island did not have mutual aid agreements with the Power companies in effect, and hired a small contractor to carry out the task of preliminary planning for the thousands of mutual aid workers that would have been initially deployed under normal circumstances.[166]

It is not an easy task to consider additional predeployed staff and assets in excess of what is normally done, which can be substantial if all systems are working as required. The process must start well before any disasters are expected. And some, for example, the location and the use of rotary wing aircraft for many treatment, assessment, critical supply, and evacuation needs require hard negotiations, more money than most agencies can apply, and no small amounts of good fortune. But often, the expenditures pay off in a more effective response. But as previously noted, the growing availability of drones will alter these response issues.

The error of assuming FEMA will "bail out" state and local emergency management agencies who are not themselves well prepared for a disaster

Even an entry level Emergency Manager at the federal, state, or local level knows that FEMA has neither been designed to, nor has been funded to, address a major portion of potential damages an area may experience from a major disaster, for example, a serious, multistate hurricane. Much of the criticism of "FEMA" is unfair to the extent that, as we will stress, much of FEMA is bewilderingly complex, and only grows more so amidst the constant administrative changes made by each ensuing Administration putting its mark on FEMA; either changing it for little or no positive effect or significantly altering its funding, too often in terms of austerity, as has been the trend in recent years. FEMA, as a federal agency, has its budget and staffing levels set through interplay of Congressional funding priorities and Presidential Administrative authorities as most other federal agencies. It is also subject to huge pressures exerted quietly, but forcefully by the many powerful contractors who have worked with FEMA extensively. But FEMA differs from other federal agencies since its own, annual operating budget ($13.9 Billion in 2018) and core staff is typically dwarfed by the flow through money it spends on state and local areas' disasters under the Stafford Act and other funding authorities. By mid-2018, over $120 Billion had been appropriate by Congress in supplemental funding for hurricanes Harvey and Irma in Texas, and the unprecedented California wildfires. As noted, it is clear that FEMA has not been designed to address more than a portion of overall damages. In 2017 the National Oceanic and Atmospheric Administration estimated cumulative damages over $300 Billion alone in the various disasters that year. Early estimates of cumulative damages from just Hurricane Maria in 2017 in Puerto Rico may exceed $100 Billion, but as of early 2018 only over $12 Billion had been obligated for FEMA's response and recovery missions in Maria. This is not to say that FEMA's disaster response and early recovery activities cannot be effective and extremely helpful, these can include: cash payments to homeowners, long-term low-interest home/small business rebuilding loans from the Small Business Administration, assistance with damaged water treatment and water purification facilities, cash payments for assistance with funeral expenses, early debris removal and street openings, provision of scores of generators for hospitals, public buildings and even key businesses, public infrastructure repair and rebuilding, public health and medical services, and many other key assistance functions.

FEMA needs to be complex by its nature and by what it needs to do in assisting state and local areas in responding to and recovering from disasters. It also conducts training, educating, exercising, and even funding these areas to effectively mitigate future disasters. As examples of this complexity and how FEMA has struggled to address its ongoing staff deficiencies, by its most recent estimate FEMA is 37% understaffed. Under the Stafford Act FEMA has the authority to augment its permanent staff when needed. Additionally, FEMA selects employees from two workforce components—the Surge Capacity Force, which is staff "borrowed" from other federal agencies to deploy during disasters but who are rarely trained or credentialed for the jobs they are deployed to do. The second workforce is referred to as the FEMA Corps., a group of intermittent federal employees who also have to be trained. According to a recent GAO Report,[167] 54% of staff FEMA deployed in 2017 was not adequately trained in the capacities they were serving in. In addition to these staff, FEMA can hire contractors and local individuals during disasters. As a general rule, FEMA nonpermanent staff are frequently poorly trained, usually severely underpaid, and often called upon to deploy more than they would want to, which has led to serious retention difficulties. For comparison, a FEMA Corps. staffer serving in the critical role of an Operations Manager will be making in the area of $30 an hour, while a FEMA permanent staffer serving in the same role may make an approximate salary level of a GS 14 or 15 in the range of $65–80 per hour. And the FEMA Corps. staff will require intense training to reach the level attained by most FEMA professional staff.

As large as the damage the nation has experienced from hurricanes, wildfires, and floods in recent years, global climate change is likely to present this nation, and the world, with larger and larger amounts of damage. But FEMA's budget and related staffing has experienced serious "boom and bust" cycles since its inception and has not been equal to the serious challenges experienced in the two-year disaster season of 2017 and 2018. This same GAO report concluded that FEMA generally carried out its duties as expected when responding within the continental US, to Hurricanes Harvey and Irma and the California wildfires, but it found that FEMA was not ready for what Hurricane Maria did to Puerto Rico. "They were completely overwhelmed from a workforce standpoint," said Chris Currie, the GAO director for emergency management issues and leader of the audit. "Once Maria hit, their staff resources were pretty exhausted. Their other commodities and resources were exhausted." In perspective, it could be added that Puerto Rico was low on food, water, tarps, cots, and other supplies as Maria came in.[168]

Though local and state emergency management and related resources are normally to be responsible for first response activities in most disasters, this is not always an effective nor a realistic posture in impoverished areas, such as New Orleans before Katrina in 2005, and Puerto Rico in Hurricane Maria in 2017. There are other examples. FEMA had to assume that role in Puerto Rico early in the response, but should not have been surprised at the devastation and lack of initial response because of the power of the incoming storm, the widespread and evident poverty, and the lack of adequate infrastructure that have affected Puerto Rico consistently, for decades, if not for most of its existence as a US possession. Though they are the exception rather than the rule, there are times and places when a predeployment of say NDMS DMAT elements, adequate local emergency management assessment staff or first response staffing is not only an appropriate use of FEMA mission assignments, but a required one. And again, we must stress that it appears that FEMA frequently has had among the highest percentages of political appointees to professional staff, many politically based contractors and subcontractors, as well as sketchy history regarding consistency of outcomes.

Errors in decision making resulting from inadequate or false information early in a disaster

Anyone who has served in a catastrophic hurricane has heard (though not necessarily read) alleged reports from a local sheriff or other enforcement personnel finding "bodies washed up on the shore." No doubt that has happened somewhere. And the worst hurricanes death counts such as the New York's Long Island Express in 1938 that killed up to 700, or the 1928 Okeechobee Hurricane in Florida that killed as many as 3000 or the Hurricane that swept across Galveston, Texas in 1900 killing between 8000 and 12,000 probably did include some instances of "bodies on the shore," one way or another. The reality of bodies washing onto shore presumably as part of a huge and deadly hurricane is that it has not happened for decades, if ever, in the last 75 years or so. Following the "dead body" theme, after the 6.0 magnitude earthquake in 2014 in South Naples, California, particularly disturbing "Hashtag hijacking" took place. For much of the initial days of the earthquake response, a significant portion of the tweets on the most popular response hashtags contained graphic pictures of dead bodies, but from unrelated events.[169] This further complicated the lives of response personnel.

There are so many significant, verified instances of rumors being spread during disasters, especially with the enhanced use of social media that FEMA opened up a rumor control website during Hurricane Florence in 2018. But the problem of rumors, particularly in the earlier stages of large disasters remains an important issue that needs to be addressed in every situation in which it appears. The era of wide internet access and various social media outlets have only intensified the problem, though recent technologies have also proved newer and powerful methods of dealing with this traditional problem. Though our emphasis here is on minimizing decision making errors because of rumors, and inaccurate reports, it is clear that rumors can also have serious effects on the population at risk during disasters and can seriously hinder the effectiveness of response and recovery actions. For example, in the Boston Marathon Bombing in 2013, shortly after the scene was secured there were media stories giving two incorrect names for the two suspects. Social media hysteria and the provision of incorrect information quickly made its way into the mass media. Fortunately, the investigation still came up with the two suspects that were eventually correctly identified.

Don't make the mistake of disregarding that recovery and mitigation are usually more complex and costlier than the relatively short period of response

During this last decade or more, FEMA has placed less emphasis and has received less funding for the recovery and particularly the mitigation aspects of its overall mission. Though this has occurred, it does not reduce the pressure on the Emergency Managers at federal, state, or local levels keep uppermost in their minds that recovery and mitigation are often more complex, involve more stakeholders, and cost more than response. Mitigation is one of those aspects that can save a lot in the future for "just a little bit" of thought and extra resources upfront, in a disaster response. It can easily be disregarded with little organizational or political costs to bear by the Emergency Manager. Only later, will the results, often costly and sometimes even deadly, be evident. During the second World trade Center response during 911, if the CDC's admonitions to wear N-95s or even respirators when working on or near the pile were stressed with the first responders and even in the media, the growing list of cancer and pulmonary-related deaths that are occurring 18 or more years after the event would be nowhere near as extensive. It was too easy just to go on searching for bodies/parts/particles, rendering medical assistance to the first responders and to do all of the other mortuary-related missions, getting caught up in the moment and the history, and not stressing the obvious and serious effects of not paying attention to the serious breathing issues. According to disaster expert, Scott Gabriel Knowles.[170] Shorting FEMA mitigation funding is a similar fault, making it a bit easier, and cheaper today, but at huge costs later.

> ...failure of imagination is ours if we continue to shortchange the Federal Emergency Management Agency and limit its activity mostly to disaster responses. Or ignore the deterioration of important infrastructure like dams and levees. Or assume that everyone can cope with a disaster regardless of age or income. Research shows that in these cases, long-term thinking and planning can save lives and dollars....Event thinking is cheaper, and it relieves us of the moral burdens....But it is counterproductive. What we should be doing is building a culture of long-term preparedness, based on scientific reality. The clear counterpoint is to acknowledge that a disaster is a process that reveals our values as a society. A disaster's beginning and end might be obscure, but that doesn't excuse us from figuring out the causes and the true costs. Surely our thinking about disasters should be as complex as the societies they disrupt. We need to understand that disaster is slow.

Knowles' words are insightful and emphasize the realities of FEMA's being funded and organized by Congress and regulated and directed by the Executive Branch in such a way that longer term mitigation activities are lower priority. This also recognizes that being "buried" in US DHS is another reality that FEMA staff must contend with as well as being perennially underfunded and understaffed with bewilderingly complex regulations and program priorities that seem to change frequently. Based on all of this, how are FEMA Emergency Managers able to put the proper emphasis on recovery and mitigation that would minimize resource use as well as maximizing future avoidance of some negative outcomes? There is no clear-cut answer, unfortunately. It leaves the Emergency Manager trying to do "the right thing" for the present and the future, but with little appreciation from the political leaders of the agency or even much knowledge of the benefits of attempting to viewing disasters as long-term events. Congress would need to act to fund wise recovery and mitigation actions such as reviewing and addressing levees, dams, valuable wetlands, and estuaries lining the ocean shores to mitigate intense flooding, national building codes for new construction that outlaws certain types of construction in vulnerable areas while requiring "hardened" construction and even retrofitting in others. So far, this is not happening to any significant measure.

The issue of longer term planning for more effective recovery and mitigation for state and local emergency management officials will tend to be more sensitive to these issues, as they will have to address the issues on the ground in their communities, and the costly and sometimes even deadly results of poor or nonexistent planning. But even this is complicated by the fact that some of the states whose Emergency Managers are well attuned to these issues, while their state legislatures are denying the reality of global climate change. The upshot is that the Emergency Manager also has to be a somewhat skilled politician if he or she is to effectively perform the job. And even then, success is by no means assured when dealing with recovery and mitigation. An excellent, recently updated (2014) FEMA report on the costs and benefits of mitigation can be downloaded at https://www.fema.gov/media-library/assets/documents/3459.[171]

The error of ignoring effective public relations because it is someone else's job

Virtually all federal, state, and local emergency management agencies as well as the ESF agencies have separate public relations offices, although of varying sizes and effectiveness. But if an Emergency Manager is being candid about public relations most will likely say "we can do without it," and the attention of the media, upper level managers or political appointees that it can attract. And, of course as previously addressed, even accurate public relations can cause the rise of inaccurate rumors and the mass panic or misunderstanding they can cause. And, in an individual perspective, an Emergency Manager has little to gain from even accurate and laudatory public relations efforts and certainly lots to lose in a variety of ways, from complicating response to spending time addressing issues raised, accurately or erroneously. And, in this day of growing amounts of data being generated, stored, and extracted, any public relation pieces are likely to open the agency to sometimes even more crushing data requests. In short, the Emergency Manager has little if anything to gain from good public relations or mass media coverage, but potentially quite a bit to lose despite great potential benefits imparted to the agency. This does not address the obvious and necessary public service messages before and during disasters that help keep the population safe. The only way an Emergency Manager can hope to mount an effective digital media messaging system is to hire or contract a professional for the job. Having someone on staff or a volunteer who is active on one of the social media platforms, but is not a skilled public relations expert is asking for trouble.

But there is another side to all of this. The Department of Defense and the HHS Centers for Disease Control have large, effective and active public relations offices, among other federal agencies. Their various functions tend to be well funded, and almost self-fulfilling regarding good stories leading to stronger, well-funded agencies regardless of the Administration and are generally respected and more or less protected in Congress. But in major or even minor events rumors, accurate, and even positive public relations stories and news coverage in general can always spin out of control for a variety of reasons, almost none of which are under the control of the federal, state, or local Emergency Manager.

So, what is a realistic, mission-oriented but not necessarily foolishly "bureaucratic risk taking" Emergency Manager to do. The problem faced by Emergency Managers is that providing adequate to good mass and social media coverage of a disaster or an emergency is the job of public relations specialists, and these staff are not necessarily well attuned to some of the difficulties that some types of even high-quality coverage can cause an Emergency Manager. The simple fact is that any public information causes questions, potential conflicts and other issues, and is the main reason the Emergency Managers can be hesitant about public relations, whether that is a wise, overall course or not. For example, quickly providing addresses for the various types of shelters and disaster centers that occur in the recovery, when even small mistakes are made, huge amounts of time and bad feelings (and perhaps Congressional Inquiries) can accompany what is basically a good job by public relations staff, but not such a good job for emergency management. And when public relations staff as well as an upper management office clamoring for "More data" valuable time is easily spent on non-mission-oriented activities (and that can be quite an understatement). Unfortunately, there is no uniform answer but to say "be careful, really careful."

The Emergency Manager can provide quick, correct, and extensive information to any media or agency upper management and public information office inquiries. And if the Emergency Manager at any level can have at least one or even two staff members to do nothing but dealing with data requests and dissemination, that is at least a start. Of course, in the largest disasters and emergencies that occur, having a few data staff will not necessarily be a limiting factor, they might even be borrowed from FEMA's Surge Unit. These are the types of relationships that need to be addressed before any disaster, for the most part, but if that is not possible it can still "be patched together," with the Emergency Manager maintaining constant awareness that those making data requests have different organizational needs than he or she does, and act accordingly.

Haddock et al.[2] take an advanced and wise viewpoint, stressing that the role of social medial in operations and decision making continues to evolve as it is increasingly used for communications in preparedness, response, and recovery. They feel that the changing shape of homeland security and emergency management in the coming years, and to a great extent in the present, will demand that communications take a much more significant role in all emergency operations and programming:

> Incorporating digital and social medial forms and functions into communications plans and strategies and adapting to new technologies to glean the data generated by social media users will be the order of the day for all emergency management agencies. Emergency management and homeland security officials can no longer avoid communicating with the medial and the public. Emergency agencies must accept the expanded and changing role of communications in all four phases of emergency management and embrace it as a valuable tool in the needs of the public. The old boxing cliché "You can run but you can't hide," unfortunately grows more true each day as we apply it to all forms of communication during disasters and emergencies

Don't make the error of not addressing, economic, health status/ disability, linguistic, racial, and ethnic differences in response and recovery

By now it is a truism, often repeated in this text, to observe that the poorest residents in an area generally live in the lowest, most flood prone areas, in the most weather-damage prone housing, lack self-protections, and often have medical issues in addition to those related to poverty. These residents are also likely to lack the resources to fix their flood or wind damaged homes (if they are fortunate enough to own them); they also probably may not have adequate insurance or many liquid resources enabling them to rationally choose between evacuating an area, temporarily staying in a shelter or deciding to "ride out the storm," which often becomes the default option for the poor (Fig. 2).

FIG. 2   Often there is more consistent care for the dead than the living.

One of the worst emergency management errors made at state and local levels immediately after Hurricane Katrina was in not recognizing the inability of many Ninth Ward residents from having the practical ability to evacuate; or to take advantage of FEMA/SBA home repair loans in a timely manner. Hurricane Katrina demonstrated that many had been marginalized by the more extreme social conditions many decades before the storm. Katrina also made it evident that table top drills, media stories, technical expertise, and science-based advance knowledge about known barriers to evacuation and other measures were, in fact, insufficient to prepared the residents of the Gulf Coast for a severe hurricane.[67]

A 2019 Report, *The Storm after the Storm*[172] issued by Disability Rights North Carolina found that the response to Hurricane Florence by public officials was inadequate. The Report found:

- Inaccessible and challenging shelter environments
- Unfit medical shelters
- Lack of important services
- Inadequate state and federal services

> The Report stated that many of the dislocations and traumas observed could be avoided if future disaster plans and recovery efforts are more inclusive and respectful of people with various disabilities and ensure their civil rights.[172]

In Puerto Rico in 2017, the failure either to gather or to be guided by these types of information no doubt led to deaths. For example, 58 of Puerto Rico's 68 hospitals lacked power and fuel directly after Hurricane Maria had passed, with some waiting for power to be restored for months; one hospital had its fuel delivered by armed guards to prevent looting.

Without power and running water, both ordinary people and medical facilities can't do tasks as simple as keeping medications refrigerated or sanitizing equipment.[173,174]

Hospitals, skilled nursing homes, and community health care centers that did not have support generators predeployed, meant that the generally old infrastructure, the lack of generator rebuild kits and the immediate available of staff to use them, and the general lack of helicopter medical assessment meant that these facilities were left to struggle on their own. This also meant that as power stopped, patients on respirators, patients needing renal dialysis, patients on wall hook ups for oxygen, and others with similar needs would be placed in situations of desperation. If either Island power were not restored (in virtually all instances, it was not) then many patients in the intensive care units of the nation's hospitals would be in danger of using their lives. Of course, nurses and others could "hand-bag" (Ventilate manually) patients that were on respirators, for example, or otherwise needing assistance, but when they did so they were not available to do anything else until the patients benefited by power or died. Of course patients needing surgery or other complicated services were further endangered. Surprisingly, these issues, though covered in a variety of mass news outlets and institutes, were not covered in FEMA's Hurricane Maria 2017 After-Action Report and neither were the projected reasons for the nearly 3000 Hurricane related deaths (Fig. 3).[175]

Six months after Hurricane Maria the Island's health care system was still slowly recovering. For example, as a very poor area the 20 federal Community Health Centers at 93 urban and rural sites served 353,172 patients, over one in ten of the island's population, and provided 12.5

FIG. 3   Planning is a huge component of proper emergency management decision making.

million patient visits in 2016. Six months later the centers have seen improvements but still face many challenges including the approximately 10% that still had limited or no grid power.[176]

An effective Emergency Manager cannot disregard population attributes in key preparedness decisions, and certainly not in response and recovery decision making. But it is relatively easy to do so in initial phases of response, especially. It goes without saying that a "group inventory" needs to be accomplished and its constantly changing composition needs to be uppermost in the minds of the local and state emergency manage. Local city planning, regional planning, various public and private real estate groups, and public health departments conduct such assessments as necessary to their core missions. Fortunately, technology merging the use of Artificial Intelligence to the massive, new data bases that are increasingly becoming available, may soon be another tool in the Emergency Manager's preparation, response, recovery and mitigation arsenal. Unfortunately, promise of these valuable tools may still be developing.[219]

Place the sentence that starts with "Religious organizations" after the above sentence that ends with "to their core mission" so that there is no paragraph break. Religious organizations and some social organizations, such as the VFW, the Moose, the American Legion, the Shriners, and so on, and even local business or homeowners' associations also conduct these assessments and can supply powerful local networks if necessary. These results must be accessed by Emergency Managers and, quite literally should be mapped. Visual representations of "Who lives where" are basic to effective community response to any level of disaster or emergency. Knowledge of the community at this level is a basic part of the City Planning curricula and is a major reason why these degrees are much prized in FEMA. As a graphic example of the failure of FEMA's attempts to assist poor residents after Hurricane Harvey in Houston, many thousands of poorer residents remain homeless, some being inexplicably turned down for many forms of relief. Inexplicably nearly 27% of Latinos, 20% of blacks, and 11% of whites reported their homes were still unsafe to live in up to a year later. Most of these residents were clustered in the poorer sections of Houston. A $5 Billion Community Block Grant fund approved by FEMA for Houston had not been dispersed by Fall of 2018.[177] Emergency Managers must be aware not just of differing population characteristics but must be especially sensitive of those with special needs, and how these affect not just the response and recovery decisions, but how these population differences affect individual and group response to disasters and emergency conditions.

It is much too late, during a serious disaster, to learn that there may be religious restrictions complicating the work of victims and Emergency Managers of different sexes, for example, especially if there are language difficulties further complicating effective cooperation; demographic types of conflicts can seriously damage a well-planned response or recovery, but can pose almost no problems at all if they are dealt with in a timely manner (i.e., before disasters).

The problem of ignoring individual predispositions, differences, and attributes when making emergency management decisions

Every Emergency Manager has observed other Emergency Managers or other individuals that almost always "get it right," whose decision making processes and data and fact gathering seem to lead to helpful outcomes. We can call this ability "emotional maturity," wisdom, or just plain luck, but it is evident. Much of what will follow has been summarized in Chapter 1 regarding various aspects of decision making. Again, the major theme here is that an effective Emergency Manager should develop the habit of being aware of his or her decision making process and of the need to constantly try to improve it, still recognizing that this is usually very difficult. And also, as Kahneman has stressed, almost ironically, we can see the flaws as well as the strengths in the decision making processes of others much easier than we can see or change our own. Again, the wise Emergency Managers can learn from others.

Before we can begin to focus on the attributes and the effects that others have on our individual decision making processes, we should have conducted some relatively quick fact gathering to put the overall disaster in perspective. Initial steps should involve information and perspective gathering. When in the field, the local fire department, post office, police department, and even the libraries, especially the selections covering past disasters, can be very helpful, as can colleagues or others who have been faced with similar situations can be called, or better yet, visited, if at all possible. FEMA's After-Action Reports (usually), GAO studies, Congressional Research Service reports, institute studies such as those conducted by the Pew Charitable Trust, local newspapers, mass and social media sources, and even the *New York Times* or *The Wall Street Journal* are all useful in developing background information and perspectives before you even approach one of the operations centers. Some might "push back" at the inclusion of the *New York Times*, but The *New York Times* covers catastrophic disasters as local and national events, as being politically selective, but will lose an excellent source if they let that feeling guide them. The NYT brings in history, politics, and even references to any in point research. With a decent smartphone much of this is just a few calls and then a few clicks away. Of course, this may immediately immerse the Emergency Manager in perhaps too much social media coverage, where it is often difficult to separate the correct and useful from that which is neither. Some may believe there is no need to act like a researcher, but so much is available so quickly, it is foolish not to at least skim some of what is out there. Be assured that many political actors have their staffs do this for them, early in any significant disaster and will probably also be enlisting the help of various Emergency Managers, willing or unwilling, in this task.

In the middle of a large and complicated response, or even in one less complex, the tendency is to look at the problems that need to be addressed and the missions that will have to be accomplished to that end. But a brief consideration of those around us, as complex individuals and groups, is good place to avoid accepting what could be negative realities just because we have not considered them. We can start with people around us who get along, and those who may not. We can't assume rational decision making from those on the decision making groups around us. Old angers die hard, old friendships can remain strong. Quick scans of those around us who may have some of their decisions skewed in any direction because of past events are important. Wise and experienced Emergency Managers will usually do this "people scan" instinctively and this of course includes a scan of our won past and present relations with individuals and groups we now have to deal with; but as we hope to have conveyed, it is best not to assume flawless reasoning on our parts, or anyone else's—ever. As an example of how individual complexities can hamper good decision making the following case is presented as follows:

One of the authors was a local plan commission chairman and had a complex housing project before him, that included 7 acres of extremely rare, "specimen tanglewood oak trees" one of the largest such groupings in the nations. To save the trees as part of the project, the Plan Commission had to make a positive recommendation to the Board of Trustees, which had the final authority and deeply wanted to save the trees. However, five of the members had a deep dislike of the project sponsor, a local housing developer, but also wanted to save the trees. The planning staff prepared an accurate paper, detailing all issues. The final vote was against the project, with four of the five "angry" commissioners voting against the project despite their desires to save the trees, stating their hopes that the Trustees could somehow skirt the rules and get the project approved with the trees stand intact. Apparently, the anger or dislike of the land developer swamped the member's ability to logically address the project. Emergency Managers do not often face such open displays of bias against an individual regarding disaster response and recovery. In fact, a wiser Plan Commission Chairman would have spent more time "working" the members who looked likely to vote against the project, but he didn't, so he also shares good part of the blame for the loss of the specimen tangle wood oak trees. Expecting people to overcome emotional predispositions just because a project is important, can be irrational. Aberrant behavior which is emotionally tinged and quite destructive can always surface.

A wise Emergency Manager is always mindful of the rare, but always possible cases of "emotional poisoning."

Fortunately, personal biases against other individuals or even certain courses of action, often can be foreseen, if the Emergency Manager is alert to small signs which can indicate these types of reactions. However, emotional predispositions against individuals or a specific course of action when manifested by some high-status or otherwise powerful people can make some difficult situations even more difficult. We've already talked about the problem of dealing with many doctors, powerful lawyers, high status consultants, some top level city, state or federal managers, and similar high-status people can sometimes be even more difficult to deal with. Mixing these problems with emotional dispositions only makes the job of the Emergency Manger more difficult, but it does happen, and not infrequently.

Although it has almost gotten to be an administrative cliché with individuals categorizing a decision making path they may not like as "Confirmation Bias" (Really frequently used) or "Loss aversion," or scores of others, groundbreaking psychologists and behavioral economist have long since recognized and proved patterns of biased "rationality" that affect us all, to varying degrees. These are merely putting formal labels on biases—sometimes ones that we are aware of intuitively, and others which it helps having a label or a name for. We mention this at the risk of redundancy, but the importance of constantly being aware of emotional biases and the effects of "noise" in ourselves and others cannot be stated enough, if improved decision making is what we are attempting to achieve. Wikipedia may not necessarily be a definitive source of commonly accepted cognitive biases, but it lists them and provides a good set of references and further reading. The list of cognitive biases is available at https://en.wikipedia.org/wiki/List_of_cognitive_biases.[178] The point is not necessarily to review them all, though that in itself makes for a relatively short read, just a few pages of succinct summary of the various biases that have been studied and named. This is in line with the ongoing theme of this book, that the wise Emergency Manager considers decision making an important process to be aware of, and a process that we should try and improve to the extent that we can.

It is important to review what Kahneman et al.[26] have found and written about how unreliable human decision makers can be. Again, the authors stress, it is not only our own decision making that is unreliable, but that of those individuals around us. Although, as noted before, it is ironic, we are often more effective in observing the potential cognitive biases and other problems affecting those whose decision making we are observing than in unearthing and recognizing our own cognitive biases. We could go on.

Emergency managers disregard political considerations at their peril

Emergency Managers generally come face to face with political considerations in one of two different ways. The first is the most direct in that most local, state, and federal emergency management agencies are run by a manager or director who has been politically appointed. Sometimes this individual has emergency management experience or training, but often does not, and is dependent on the permanent professional staff to provide necessary guidance. This fact can cause real difficulties at many levels but is the reality that must be dealt with. The *Los Angeles Times* addressed this in an in-depth article about FEMA political appointees.[179] MIchael Brown previously discussed in more detail, was one of at least 5 other top officials at FEMA who were appointed under George Bush directly before the Katrina Hurricane agency mishaps. In fact, Jim Albaugh, the FEMA Director who was replaced by Brown was himself appointed with no significant disaster or emergency experience and was perhaps even more responsible for later agency activities, as he left quite a number of FEMA's Administrative directorates vacant or severely understaffed at the higher levels, after former Director James Lee Witt left. As a counterpoint to Albaugh and Brown, FEMA Director Witt has received much praise, not the least of which was in staffing his upper level management with experienced, highly qualified Emergency Managers. In fact a year after Katrina the head of a labor organization representing FEMA employees wrote, ".... Emergency Managers at FEMA have been supplanted on the job by politically connected contractors and by novice employees with little or no background or knowledge (of disasters)." The problem is that disaster and emergency issues can be easily grasped (but certainly not mastered) by those with no emergency management background. So, the threat of less than desirable FEMA or other emergency management leadership remains a constant one.

The best response that an Emergency Manager can make to the politically appointed agency leadership is to place himself or herself "in the shoes" of the political appointees and realize some of the pressures they experience from their political superiors, from the mass media and now even the social media, and do the best that can be done with the tools at hand. The *Washington Post*[182] made a number of wise suggestions that permanent, upper level federal employees (Senior Executive Service) could do to smooth the transition from one group of political appointees being replaced by another group, as the Presidential Administration changes. These comments came from John Pagutta, a 34-year veteran of the federal government who spent years watching the transition process unfold under a variety of administrations. His comments are virtually applicable to state and local governments and are valuable to consider, wherever we are on the management chain:

With the start of President ---'s (last term) many political appointees have already departed or will soon be leaving—and their replacements slowly but surely will be arriving to take over key management positions. This is always an unsettling time for career executives who must manage operations during the hiatus and then adjust to any changes that inevitably will take place. However, it is also an opportunity for you to form a productive working relationship with the new political appointees.... To assure a positive and productive start, here are some of the pitfalls:

- *Don't begin your first conversation with a new appointee by preemptively telling the individual what needs to be done immediately.*
- *Avoid responding to new ideas with, "That's not going to work because..." Even if the ideas have some flaws, such as being in conflict with existing rules, think about alternatives you might suggest.*
- *Resist the temptation to criticize your agency's former political team or to list all the terrible things the new appointee's predecessor did. You'll only make them wonder what you might say about them when they leave.*
- *Don't hide problems. Bad news doesn't get better with age, and being upfront about problems will sting less than if you wait until the problems grow or become public.*
- *Curb any thoughts about being a yes man or yes woman. For example, don't agree to do something if you know that you don't have the ability to follow through.*
- *If you simply cannot or will not accept changes a new appointee may legitimately wish to make it may be time to consider a job change.*
- *What can you do to help yourself, your new political boss and your agency succeed? Here are some of the affirmative steps that you can make—*
- *Learn all you can about any new priorities and goals that your new boss may have for the organization. Some online research before the appointee even arrives can be helpful.*
- *Be open to and ready for change; that's what it's all about.*
- *Anticipate questions and have constructive answers ready. Hint: Simply telling a new appointee that "we've always done it that way" is not a constructive answer.*
- *Develop suggestions on how to best address any external criticisms. For example, if the organization has been publicly criticized for unresponsiveness, be ready with viable recommendations.*
- *Assume the best about the new team until your proven wrong.*
- *Focus on the end result—the job the organization is there to do for the American public. Work backward from that objective and figure out the best way to get there together.*

For career senior executives, the law provides for a 120-day moratorium on involuntary reassignments after the appointment of a new agency head or the appointment of a new direct supervisor who is a political appointee. (It is) recommended that career executives use this honeymoon period not only to get to know the new appointees, but also to make sure they have a chance to see that you are a valuable member of their team.

An additional critical issue in dealing with political leadership, though not one included before, is that too often newly appointed political leadership can have an unrealistic assessment of their own knowledge and judgment in the area in which they now are among the top managers, after all, they are in charge, and have been selected by even higher political leadership. Sometimes this feeling of overextended abilities is over soon, withering in the cold realties that will be faced, but sometimes it can extend for the whole period of the appointee's term. This has been addressed earlier but has been readdressed here because it has been and will continue to be a difficulty that Emergency Managers will need to cope with as long as our system is constituted as it is.

The second way in which the Emergency Manager faces political considerations is that during almost any disaster or emergency situation, especially those that are large or may involve extensive potential damages or loss of life, the public has a strong stake in understanding what is being done by various agencies and/or the military on their behalf, and what the likely outcomes of the disaster or emergency situation will be. A strong public interest will be reflected in strong political interests. A difficulty implied by political interest in a set of issues is that the Emergency Manager's authority, funding, and even ability to make certain decision can be seriously affected, or they may be left totally intact by wise political leadership. And Emergency Managers can never be certain of the final courses because information in those areas will not be generally available. Though political leadership may or may not be very knowledgeable about the disaster response going on, they are wise in the ways of politics and have learned that it is often best to let Emergency Managers do what they believe to be the right thing, and then "justly" claim credit as the events unfold. Of course, they can easily blame federal, state, or local Emergency Managers if difficulties occur. Damage may flow downhill no matter what the Emergency Manager does, but as noted before, being a valued member of the team will help in achievement of the mission as well as protection of one's status (in most instances).

The model, if not the false ideal of "nonpolitical" disaster response is too often the conventional thought in too much of emergency management, though the higher levels of education that recent emergency management graduates are receiving, along with the accelerating professional "demise" of the baby-boom generation seems to be giving political concerns the stress they should always have been be receiving. As Emergency Managers have, for decades wrestled with effectively dealing with political considerations, so have urban and regional planners, a professional field that FEMA has long valued as a potential pool for staffing. Writing of a freeway placement in Cleveland, decades ago (with lessons till relevant to today's planners and Emergency Managers) Norma Krumholz and John Forester wrote:

The model of apolitical planning dies hard, in part because we have too few examples of exemplary, equity-oriented planning that weaves together professional work, political vision and organizational pragmatism. .... As they worked with community leaders or mayoral advisors, with agency staff, specially created single-issue task forces, the Cleveland planners were able to develop, a largely public, equity-oriented voice that integrated professional analysis and political initiative. They did not sacrifice professional integrity to political pressures. Instead, their work teaches us that had they not given voice to the equity-planning agenda (fairness at the neighborhood level, regarding placing the express through poor as well as more affluent neighborhoods), had they not pinpointed the vulnerability of the transit dependent, had they not actively protected tax-payers facing ill-conceived infrastructure projects, the planners would have had far less to show for their professional work. The Cleveland planning experience demonstrates the possibilities of politically astute, articulate, and effective equity-level planning practice.[183]

But all hard politics are not of the partisan, Democratic-Republican kind, nor do they have to be dealing with politically appointed agency leaders, though as we all know these occasionally can be quite nasty. Political scientists have long known, as well as those who have been involved with local government that the politics closest to the local level can sometimes cause the most personal conflicts and can draw out the strongest emotions. Small issue like where you walk your dog, park your car, decorate the front of your home or condo can evoke emotions much stronger than any might imagine because the individual is directly affected, and those doing the affecting can be seen every day, up close. It stands to reason because people either playing in local politics or being affected by them can and do see each other daily sometimes, up close at the stores, many night meetings, at kid games, bars, and restaurants. Another example is local school boards, where one of the authors served two terms and was amazed at how emotionally charged issues can be which no doubt carries over to other interactions with people on opposite sides of the issue.

But this is not the only place that "hardball" politics are played. Politics at the upper levels of local fire and police departments and even the upper levels of union management are often played almost as a "blood sport" where those on the wrong side, whichever side that is, can easily pay with their jobs or their reputations. The Emergency Manager will try and avoid these often emotion-laden entanglements, but has to realize the intensity of personal feelings that can accompany politics as it is "played" at these levels. The former Mayor Richard J. Daley of Chicago once said, "Good government is good politics." He was right, but there was so much more that he did not say. People and their emotions can be a lot more complicated than even the most difficult policy or legal issue and can cause raw emotions that can last hours, days, or even years.

And there is another level that should be considered, though not necessarily dealt with—political paybacks, both positive and negative. For example, local governments with their zoning issues, purchasing authority, and need to tax and spend money will always be subject to the rules of "clout" to use the Chicago term, at any level. Most political considerations that are tawdry or that approach illegality are not usually associated with bags of money in unmarked bills being given for voting a certain way. It is much subtler than that and usually involves something like helping a brother-in-law get a job, for example, from a developer that is doing business in a village or town, or a job with the company that has for years sold masses of salt to a suburb for spreading on the streets in snowy winters. People have to be hired, contracted with, or otherwise dealt with in various business manners, all the while getting the best prices for the best services, often with competitive bidding processes. To assume personal and political reasons are not behind some of these choices is not realistic. Of course, most villages and towns have rules covering conflicts of interest at various levels, nepotism and its many variations, but rules and laws are made to be broken. But money multiplies everything. A few years ago, one of the authors was in the top floor of a large Midwestern city skyscraper, celebrating a late-night drink with a powerful local lawyer; they were celebrating the development of a local citizens' board required on a hefty federal grant, when the lawyer responded strongly to some comment the "fed" made. He said, "Where did you come from, Chicago right, wouldn't you have figured that each of the last ten, big downtown projects were "juiced" here or there." He wasn't referring to a morning drink. Of course, all of this is not to imply that every governmental deal, small or big, particularly big, has corrupt potentials, but that they can. The Emergency Manager always has to assume that there are things going on that he or she is not aware of and will never know. This does not necessarily imply illegality, but it virtually always implies complexity.

Imagine trying to deal with both the policy considerations as well as the emotional considerations that local politics evokes in an area as already as complicated as ocean front mitigation after a big hurricane, with considerations of flood insurance and complex rebuilding issues, insurance premiums, homeowners' associations, oceanfront condominiums (and their condominium boards), and even complex FEMA rules involved. As Col. Boyd might have observed, "Learn as much as you can about every issue you may encounter and about as many models of thought as you can conceive. It is best to know as much as you can," and as many people as you can, as any politically astute individual might observe. Viewed in a wider perspective, we can sometimes wonder how we get anything done without excellent relationships at that one moment in time.

Don't expect FEMA or any federal agency to accept responsibility for security during disasters or emergencies

During a complex response to a disaster or emergency many situations will surface that will require trained, armed security. Unfortunately, adequate security for staff and vulnerable assets usually cannot be assured. What is "Everyone's concern ends up being no one's concern" too many times. There is a 58 federal agency Interagency Security Committee that coordinates efforts to protect federal employees and federal facilities, but no such group exists to coordinate security during disasters or emergencies. On one hand this may appear to be an unfair criticism. During disasters or emergencies local and state enforcement capabilities will, by definition, be stretched to capacity and will no doubt begin to request mutual aid security assets from surrounding communities, and even states in large enough events. But this takes time and is best accomplished when there is advanced information about an impending weather event or even a planned event, such as National Special Security Events. But even then, the arrival of adequate security takes time, even when requested in a timely manner (Fig. 4).

These statements are not meant to imply any serious systemic failures in the use of mutual aid systems to augment needed local security. But things happen before and during catastrophic events and even well-planned NSSE. And given that security staff come from local and state police departments, local and state police departments arriving under mutual aid agreements, from the National Guard (local and/or from adjacent states), from various federal security sources (Federal Protective Service, Secret Service, security

FIG. 4 The need for security rapidly exceeds the immediate capabilities in major events. Other solutions need to be considered.

from some federal agencies and even from contractors). Each of these security assets has its own mission and its resources allocated only to that task. For some federal agencies with no security assets of their own, but with security concerns this can present potentially serious concerns (such as the NDMS DMATs).

Don't disregard public health issues during recovery or response

Unfortunately for the Emergency Manager, it is "Pay me now, or pay me later," despite the fact that ESF #8 keeps a tight hand on the public health aspects of the response. Makingnone the less a real responsibility of emergency management. Almost any major disaster will have public health and medical system aspects to it. Major flooding and major hurricanes, which almost always include some aspects of major flooding, will include public health issues. For example, in mosquito season the intense water puddling can cause swarms of mosquitoes that can actually kill livestock by clogging their noses or by damaging their immune systems with hundreds or even thousands ofof stings. Of course, Zika and a wide variety of mosquito borne threats expand as does our trade exposure to the world. And flood water itself is subject to carrying whichever industrial, sewage, and so on, waste is in the area. Flooded pig farms and even chicken or turkey raising ranches can present huge water issues as can the failure of both water purifications plants as well as flooded sewage systems. Local public health department run the gamut from well-resourced and professional to small and mostly nonprofessional. Calling in the US Centers for Disease Control early in an event, or even before the event if public health issues are expected or have occurred before in the past, is a wise idea. Calling in public health mutual as well as EMS-type mutual aid, such as the Midwest's Mutual Aid Box Alarm System is also a smart decision providing vehicles, trained staff that can be used in a variety of ways, often even outside of their main areas of expertise, when manpower is needed.

We cannot expect things to "just work out." For example, during the WTC 911 response quite a number of responders working on or near "the pile" did not wear adequate respiration protection, and sometimes none at all. And surprisingly, some big agencies that should have known better did little to stress hard enough the CDC message of at least using N-95 protection. Unfortunately, some of the same agencies have failed to monitor and assist a growing number of former staff who have had cancers that have a high probability of being related to inhaling the toxic mixture of burned metal, burned body particles, and so many other contaminants and irritants that were in the air for weeks after the collapse of the towers. Both authors of this text were on the scene and can attest to those working on or near the pile without adequate protection. A well-prepared Emergency Manager has learned these lessons in class, in readings as well as from available SMEs. But does not fail to watch the progress of the public health response, not reaching out of his professional lane, by applying good sense to what is seen.

An example of the common-sense lessons that can be applied by nonpublic health professionals is that all flood water is to be considered seriously contaminated. And when it soaks wall board, the wallboard and the furring beneath it must all be removed, as expensive as that can be. One of the authors had an experience with this subject that was handled very well by a public health emergency manager that was still in the learning phases. They were part of an ICS 300–400 class the author was teaching. The evening before class day four was to start a water main burst in the county owned building. He had to go before the county commissioners to justify his expenditure of funds to remove, dry, and replace the wall board so quickly. Because the emergency manager had written (a new skill he was just learning) an IAP and included the removal, drying, and replacement of the wallboard as a life safety objective, the commissioner's signed off on his emergency expenditures.

Animal evacuations (really hard to do, as one would imagine), additional protections for waste pools, and a wide variety of response activities, large and small, should be considered as soon as possible, certainly before the event has started, if that is at all possible. Having more rather than less National Guard available is also always a good idea. It is widely known that FEMA's FCOs are quick response, types of people not normally geared toward public health issues, but the time spent in briefing them and seeking quick mission assignment assistance is time well spent. Jim McKay in Hurricanes Florence and Michael raise the Issue of Public Health During Disasters, Emergency Preparedness.[184]

Don't disregard the mental health of first responders or victims

During most disasters one of the last things that are on an Emergency Manager's mind is the mental health of first responders. The fact that the divorce rate of law enforcement personnel is around 75% and more than 80% of firefighters experience symptoms of mental health issues. Almost 25% of dispatchers have symptoms of Posttraumatic Stress Disorder.

Since physical safety is uppermost in the minds of Emergency Managers, it is too easy to gloss over mental health. After all, the first responders are often thought of as able to work and even save lives under the worst conditions imaginable, with no long-lasting results. This may be changing, writes Jim McKay, in Preparedness, quoting Dr. Stephen Odom, CEO of

New Vista Behavioral Health, "They hear everything and see everything, it's the worst moments in people's lives and they're supposed to be calm and handle it.... They compartmentalize it, and you can only do that for so long. Over time, it begins to take its toll." He went on, "More and more, they are beginning to talk about mental wellness now and not mental illness. Departments need to acknowledge the issue." Dr. Odom says agencies have the resources, they're just have to acknowledge...and lead the charge...There are employee assistance programs, departmental wellness people, but they have to be given permission and clout to put these things out there so people want to be part of it.[185] In Bourbonnais, Illinois in 1999, where the Chicago Bears have held summer training, a truck driver drove around a railroad crossing sign killing eleven and injuring 121. Among the dead were two little girls, who died almost instantly from the fumes, and looked like they were sleeping peacefully, clutching each other, and their teddy bears. One of the authors was on the scene with an NDMS Disaster Mortuary Team (DMORT). For the few days that the DMORT was working, the restaurants in the town would not accept payment from any federal employees who would normally not accept these types of gratuities. But in this instance the normal rules were waived amidst the shock and sadness virtually everyone in town was experiencing. The DMORT staff, normally a mission-oriented, disaster hardened crew, did not rush up to work the two young girls, preparing them to be sent to funeral homes. The Regional Administrator from ASPR respectively refused to view the bodies as might have been normal. Sometimes there is wisdom in not adding to the list of horrible things that have been seen.

As important as it is to avoid disregarding first responder mental health issues it is probably even more important to keep victims' mental health issues in mind, all during the disaster response period. A recent article in a legal journal emphasized that successful recent attempts to formalize and standardize disaster definitions and response protocol will improve efficiency, clarity, and coordination, but they may fail to consider that increased systemization may result in unintended, deleterious consequences such as subverting or distorting empathetic decision making and prosocial motivation essential to effective disaster management.[220]

Don't' assume dealing with psychopaths is rare

We have not integrated this section into others dealing with various aspects of normal Emergency Managers and citizens making decisions before, during, and after disasters. Psychopaths are, as we will estimate, relatively rare, but since their effects on situations and other people can be so strong, they deserve some special consideration. Psychologists estimate that between one and 4% of the population has a psychopathic personality. By this we simply mean individuals who have little or no conscience, and who are not restrained by the normal and accepted ideas of right and wrong. The problem for Emergency Managers (as well as for all individuals) is that these individuals often tend to cluster at the upper levels of institutions like the military, major corporations, or other organizations.

> Psychopathy is among the most difficult disorders to spot. The psychopath can appear normal, even charming. Underneath, he lacks conscience and empathy, making him manipulative, volatile and often (but by no means always) criminal. They are objects of popular fascination and clinical anguish: adult psychopathy is largely impervious to treatment, though programs are in place to treat callus, unemotional youth in hopes of preventing them from maturing into psychopaths.[186]

The authors are not suggesting that psychopaths and sociopaths (for our purposes here, the same issue) are lurking behind every desk or uniform, but that some of the same traits that can make one a successful business leader or flag rank officer can sometimes be assigned to these often charming, disarming, and skillful human manipulators. Psychologists Robert Hare and Paul Babiak looked at 203 corporate professionals and found about 4% scored sufficiently highly on an evaluation test to be evaluated for psychopathy. Hare says that this wasn't a proper random sample (claims that "10 percent of financial executives" are psychopaths are certainly false) but it's easy to see how a lack of moral scruples and indifference to other people's suffering could be beneficial if you want to get ahead in business.[187] The point of this is for Emergency Managers to be just a little extra aware of those leading high-status organizations or found in high status occupations. They didn't all get there by being super competent, nice guys.

Never forget that catastrophic disasters can significantly affect history

In the heat of a disaster response, or in the longer and often more complex recovery period it is difficult for the Emergency Manager (or anyone) to ponder the long-term effects of any disaster. But it is clear that catastrophic disasters have shaped our cities and their architecture, elevated leaders and toppled governments, and influenced the ways we think, feel, fight, unite, and pray. Dr. Lucy Jones, then a Seismologist for the US Geological Survey, led a team of over 300 scientists in an effort called "Shakeout," to try and predict the results of an expected, serious earthquake along the long, Andreas Fault in California. Shakeout predicted that a serious earthquake along the San Andreas Fault could shake for an estimated 50 seconds. By comparison, the Northridge Earthquake in 1994 shook for fifteen seconds and caused $40 Billion in damages, but without seriously affecting any dense urban areas.[120] Her book is probably another candidate for the already packed Emergency Manager's bookshelf we recommend because of the many insights Dr. Jones offers that we have seen nowhere else.

Of course, obsessing on "future" issues that may not turn out well can leave even the wisest of Emergency Managers "slower on the draw," but maybe just a bit of this is good. And even events that are serious, though not catastrophic can also leave deep social and governmental scars as well as sometimes providing unique opportunities for mitigation efforts that will lessen or maybe even seriously reduce the worst negative affects of future events. For example, the 1995 Great Heat Wave of 1995 in Chicago killed an estimated 700 at a time when the City's Heatwave Plan was just a few pages, and apparently not well circulated and certainly not well exercised. And preparedness plans were mostly ignored by the Public Health Commissioner (later removed) and even the Mayor, until the deadly heat event could no longer be ignored with hearses lined up outside of the Medical Examiner's Office. During that event FEMA operated under the previous policy of "No dough for snow, and no dough for heat." The 1995 Heat wave changed that and enabled FEMA to be a stronger participant in a wide variety of weather-related issues, from then forward. The City of Chicago immediately empowered a high-status Task Force and began drafting a Heat Wave Plan (along with a few other cities) drastically reducing potential deaths from a serious heat wave that occurred just a few years later. Chicago's Plan, annually updated, remains one of the "best." Another result was the beginning of the development, along with the State Division of Emergency Medical Services, of an advanced "hospital to hospital" communications systems, enabling hospital staffing and bed capacities to be instantaneously shared among all receiving hospitals, the EMS authorities and City and State Health Departments. This avoided ambulances speeding to hospitals already on "bypass" wasting time going to another hospital with available capacity.

Dr. Jones emphasizes that the biblical "Great Flood" and Noah's leadership immediately and directly is a shared religious disaster image that persists throughout the Hebraic and the Christian world. She also observes that the earthquake and resulting tsunami flooding that devastated the ancient Roman city of Pompeii acted to topple the prevailing Roman religion and altered how people saw themselves in the world. In her research Dr. Jones uses many other significant examples of catastrophic events having effects that well outlived the event itself. Her point is directed at Emergency Management decision making and the need to be aware that catastrophic disaster events can have effects, many of them at least becoming particularly visible at the time, which can have much strong long-term effects than those cause by the catastrophic event itself.

Don't disregard scientific research

With all of the things an Emergency Manager must master and all of the relationships that must be maintained, it seems "a bit much" to suggest that scientific research is another area to monitor "out of the corner of her eye." But she should, and it is not an easy thing to do. Science works only when its practitioners are free to argue opposing sides. In effect, when Emergency Managers watch for seismological reports concerning potential earthquake eruptions, or the path and the intensity of approaching earthquakes, or impending heat waves and so many other issues are clearly science based. Disasters leave in their wake voids of information. In L'Aquila Italy, scientists were restrained in telling the population the odds of an earthquake event, being restrained by a fear of panicking the populations that no doubt should have been panicked. When an earthquake struck among the damages were Italy's faith in its seismic scientists, normally a group of skilled professionals who were held in high regard by the public. This demonstrates the problems of mixing communications issues with issues of mass warning. In a take on the CDC communications rule: "tell the truth, all of the truth, and tell it first." When scientific findings are held back or slightly modified for any reason, there is a problem that may not be fixable if an event occurs (Ref. 120, pp. 184–85).

Money alone will rarely solve problems encountered in a disaster or re-covery period, and in many instances can actually slow down or confuse necessary actions and resource acquisition. Unfortunately, the amount of resources allocated to a disaster response or recovery is rarely, closely cor-related to expected and useful outcomes. There are just too many other variables to consider that can drive spending amounts and directions:

- the complex structure of the federal government itself, that finds federal employees a distinct minority among those tasked with operating federal programs
- the potential availability of the resources in question
- the timing of the applications of the resources,
- the location of the expected resources or the expected activity,
- the types of equipment and commodities secured,
- the talent of the staff selecting resources
- the talent of the staff distributing the equipment/commodities or providing the services
- the existence of existing or potential political issues
- the availability of talented finance, logistics, and SME staff
- the existence of appropriate transportation, security, and related staff support

We are not saying that decisions related to resource allocations are so complicated that they can't be done effectively. We are saying that re-source decision making should be made with these (and other) questions in mind. But often these decisions are made "routinely," without much studied thought. FEMA FCOs, for example, mostly run toward the active, "can do," types who will more easily flow resources out during the first few days or even weeks of an event, especially in what we all would consider "response" mode; and most pointedly if it has to do with ESF8, or other life-safety-related issues, even if the relationship is sometimes tenuous (Fig. 1).

Although it is clear that the largest expenditures will be made over time, after the direct response mode through recovery periods, which can stretch out many months and sometimes years. Given this, it would seem that the fulcrum of effective disaster expenditure needs to be in enhanced recovery and mitigation efforts at minimizing money wasting "next time," recovery. The problem, one of many, is that election cycles foster short-term thinking and mitigation is based on cost savings over time, and feature indeterminate amounts of dollars, saved over time, in response to mitigation dollars spent now, by political actors and agency leadership who will not necessarily be in power when the cost savings benefits are realized. And we are speaking in ideal terms, regarding data. For the most part, as will be discussed, most states and local areas do not even have good information about what has been spent, who has spent it, on what,

FIG. 1   Large medical operation underway.

and when in disasters. For those who are students of governments, especially Emergency Managers, his should be no surprise.

As we discussed in earlier chapters, it is too easy (and very unlikely) to assume that FEMA is the focus as well as the source of nearly all disaster reimbursements; federal disaster assistance goes well beyond FEMA's Disaster Relief Fund (DRF). Of the massive 255 Billion in inflation-adjusted federal expenditures on disasters from fiscal years 2005–2014, the DRF fund accounted for only $111 billion, according to a September 2016 report by the Government Accountability Office.[188] The remaining $144 Billion or 56% was drawn from the budgets of 17 major federal departments and agencies.[189] With these huge levels of expenditures being spent with no central coordination, in fact with little coordination of any kind, we can see why actual amounts of money do not necessarily correlate with effective outcomes.

A recurring theme in other sections that an effective Emergency Manager at any level needs as much distilled fiscal information, as quickly as possible, but too often gathering these type of useful information in response to request "from up the chain." To this extent the availability of huge and relatively quickly available "mega data" can be a mixed blessing for the Emergency Manager, but one that can be handled by merging emergency management needs "on the ground" with those requested from above. More than a little diplomacy is required, as well quick access to SMEs, not only in fiscal areas, but also in data acquisition and use, and relevant FEMA, state and local procedures and administrative requirements. In the past the knowledge that the Emergency Manager would have focused on would have been gathered through after-action reports, information from those who have worked similar issues before and from knowledgeable colleagues as well as from one's own memory banks. These are still necessary, but need to be augmented, if not sometimes supplanted by these more technical needs resulting in huge databases.

Government outsourcing: A dilemma that causes much government waste and failure

Complicating this task of spending money effectively or even deciding when spending anything is at all is appropriate, is the huge issue of government outsourcing, an ongoing dilemma that is at the heart of most US government problems of waste, inefficiency, occasional fraud, and failures. Though addressing government outsourcing may seem far afield from emergency management, we will need to spend some time on this issue because dealing with federal resources and agencies (not just FEMA) is an important and continuing part of every Emergency Manager's life, no matter how large or small the agency, the private nonprofit or for profit that employs the Emergency Managers. In order to work in the hugely complex federal government, it will be necessary for the Emergency Manager to not only understand the interplay between federal employees, state and local agencies, private nonprofit, and private business it will be necessary to be able to assist state and local governments to secure grants, contracts, or service agreements, and even to successfully apply for them, depending on which part of the structure the Emergency Manager works in.

For the most part, in major Europe nations, Japan, Canada, and other advanced nations, governmental authority for purchases and services rest directly in the hands of federal employees. As we will see in more detail, in the United States much of the actual implementation of federal services and purchasing has been outsourced by contracts and long-standing agreements to various state and local agencies, private nonprofit corporations, and for-profit corporations without enough skilled federal employees to adequately manage and monitor the huge expenditures involved. This issue affects every Emergency Manager in almost anything she or he does, but is one of the complex, often uncontrollable aspects of the job that is too often glossed over, for a variety of reasons. Journalist and expert on government at all levels, John J. DiIulio, Jr. has thought and written about the often destructive and wasteful effect of outsourcing government staffing, services, and purchasing/monitoring instead of having them performed by "sworn" federal employees.[190]

Government has grown vastly bigger and more complicated over the last half century; in 2013 it spent more than $3.5 trillion, adjusted for inflation—that was five times more than it spent in 1960. In fact, the $6 trillion budget deficits from between 2009 and 2013 exceeded the total federal spending from the period 1960 through 1966. But as DiIulio states, the biggest story about big government that matters most to the future is not at all about government finances. It is the fact that during this whole period since 1960 the overall government has increased about fivefold, but the number of federal bureaucrats in 2014 stood at less than 2 million, a couple of hundred thousand less than in 1960. Today's government now spends billions of dollars a year on Medicare, Medicaid, homeland security, emergency management, housing, environmental protection, elementary education, child welfare services, historic preservation, farm subsidies and massive food support, urban transportation, rural highways, airports, food stamps, and much, much more with laws, policies, programs, bureaucracies, and regulations on numerous matters that were not even on the federal agenda in 1960 with approximately the same amount of core staff.

Even though it now leaves no area of American life untouched, how is it that there is roughly the same number of federal employees as there were under Eisenhower back in 1960. DiIulio calls this big government "by proxy" because the fundamental truth of its size has been hidden in plain view for years. The proxy groups are state and local governments, for-profit business and their plentiful contractors, consultants, and nonprofit organizations that include schools and universities and the vast grant and loan programs that directly and indirectly benefit them (Ref. 191, p. 13–16). As they develop talents and required knowledge, the wise Emergency Manager has to become an expert in the huge and complex mix of providers of government services that augment the relatively small crew of federal civil servants managing, monitoring, and sometimes just barely watching all the programs, grants, contracts, license agreements, and purchase that now constitute this massive governmental entity that (and probably already knew) is too often wasteful and inefficient. And, despite all this, as Pressman and Wildavsky[57] remind us, the program as conceived of by Congressional intent, as managed by federal civil servants, as implemented by regulations often strongly influenced by the corporations that are being "regulated" rarely resemble what they were intended to resemble. Unfortunately, as we will discuss, FEMA is part of this structure and too often typifies many of its worst aspects. In fact "Big" government is most often not even federal.

"Big Government" in America is a Washington led government by *proxy*. This fundamental truth about how big government in America really works has been hidden in plain sight for decades. Big intergovernmental proxies are state and local governments, for-profit business, and non-profit organizations (Ref. 191, p. 16).

The proxies for big government (state and local governments, for-profit businesses, nonprofit organizations)

State and local government proxies

More than two dozen federal departments and agencies spend a combined total of more than $600 Billion on more than 200 intergovernmental grant programs for state and local governments. Adjusted for inflation, between 1960 and 2012 federal grants in aid to state and local government have increased tenfold (Ref. 191, pp. 17, 18 citing Ref. 193), Ref. 95, p. 2. As a last point it should be mentioned that the argument over big government and small government is a very old one, dating back at least, to the faceoffs that former President Franklin D. Roosevelt had with Herbert Hoover, the departing President in the midst of the Great Depression.[249]

For-profit business proxies

The federal government has spent more than $500 Billion a year on contracts with for-profit firms, many of which have the federal government as their sole source of revenue. In 2012 the Department of Defense added an additional $350 Billion to its 800,000 defense contractors (Ref. 191, pp. 17, 18 citing Ref. 192, US Government Accountability Office,[193] US Government Accountability Office[194]). But despite this complexity, in fact to a great degree on account of it. Michael Lewis points out that the basic function of the federal government is manage risks that we as citizens, private businesses, nonprofit organizations as well as state and local governments cannot handle effectively and efficiently by ourselves, even acting in groups. Some of the risks are easy to imagine: a financial crisis, a hurricane or a tornado, or even a terrorist attack (p. 25 of the Fifth Risk, in the References and written by Michael Lewis). In addition to the all-pervasive outsourcing problem, federal agencies are powerfully constrained by limitations on the ability of their managers to buy and sell products or even hire and fire people on the basis of what best serves the efficiently and productivity of the agency.[51]

Nonprofit organization proxies

The nonprofit sector encompasses about 1.6 million organizations. Forty percent of the nonprofits file IRS reports and have about $2 Trillion in resources, roughly a third of which comes from government sources. In that same year government entered into 35,000 contracts and grants with about 56,000 nonprofit organizations and paid $137 Billion for service (Ref. 191, pp. 17, 18 citing Ref. 195, see also Sherlock[247]). Congress routinely emphasizes preventing corruption over achieving efficiency. In both the court of public opinion and in courts of law, the best defense for both elected and appointed officials when such things happens is to blame the proxies for not doing their jobs well enough. The state and local governments, private nonprofit, as well as for-profit businesses have learned this lesson and seek ever more detailed grants and contracts, ever more precise reporting requirement. The federal bureaucrats are quick to oblige them. Over time the system become ever more rule bound and ever less concerned about performance, results, cost-effectiveness, and efficiency gains. The grantees or contractors, the agencies, and the congressional oversight committees each find a refuge in red tape, as DiIulio wisely observes (Ref. 191, pp. 39) and Congress with both political parties using ideologically tinged arguments, content to berate wasteful federal bureaucrats. As an ongoing complexity regarding private nonprofit organizations, 501(c)3 types of nongovernmental organizations cannot politically lobby, examples of these types of organizations are religious organizations, charities, educational organizations and even scientific, and amateur athletic associations. These type of organizations can receive government grants/contracts to participate in various stages of a disaster. They cannot politically lobby federal agencies, though they can seek federal grants and contracts. In contrast, 501(c)(4) organizations can conduct politically lobbying, though that cannot be their sole or main function. Examples of these types of NGOs are social welfare organizations such as civic or neighborhoods associations. Federal grants to these types of organizations are at least theoretically possible, though unlikely in most instances. Of course Emergency Managers that must deal with these types of NGs must have ongoing legal (and maybe political) advice before, during and after disasters.

Unfortunately, most Americans do not believe that the percentage of our gross national product spent on government services is not that far from that spent in other advanced nations. This should be no surprise because the United States is mostly alone among advanced nations in "privatizing" the provision of many government services, many of which are produced by contractors who often have a strong role in the development of "on the ground policies" that would be otherwise developed by federal employees in most other advanced nations.[196] When these policies fail, as they did in 2018 when Veterans Administration IT contractors made errors that resulted in nearly 100,000 beneficiaries not getting their monthly educational stipend checks, the "Agency" was severely criticized. The IT folks were not even named as contractors. A better way to get an understanding of federal expenditures is to realize that most federal laws, in whichever policy domain, authorize Washington to do one of four things: (1) Pay subsidies to particular groups and organizations in society, (2) transfer money to statist and local government, (3) award grants or contracts for nonprofit firms and nonprofit organizations, and (4) devise and enforce regulations on the society and the economy. DiIulio calls it the Leviathan by proxy, a uniquely American, superficially antistatist from of big government that has entered into every nook and cranny of both public and private life. Our Big Government dressed as state or local government, private enterprise, or civil society is still big government so that growth in this American "state" is much harder to restrain and its performance ills are much harder to diagnose and cure, than they would be in a big government more directly administered by federal bureaucrats themselves, as most American erroneously assume they do (Ref. 191, pp. 41, 42).

Federal contractors

Despite their huge effects on US governmental processes, information on overall types, skill sets, credentials, salaries, and benefits for federal contractors are not kept by the US government (fortunately, Neal Gordon,[197] has estimated that about 40%, or 3.7 million of the federal government's 9.1 million active employees including contractors, postal workers, and grant employees are contractors). Gordon based his estimates on work done by Paull Light.[198]

Light tracks the growth of contractors in the US workforce and has long voiced concerns over what he calls "the shadow government." In the Report, Light writes that contract employees "work in a hidden bureaucratic pyramid." While presidential candidates campaign on promises to cut the size of government their proposal never mention the growing size of the contractor workforce. He warned that the blended workforce may have grown so large and poorly sorted that it has actually become a threat to the very liberty it is supposed to protect. He concludes that:

> It may have become so complex that Congress and the president simply cannot know whether this blended workforce puts the right employees in the right place at the right price with the highest performance and fullest accountability.... in his discussion of the various systemic "pressures" compelling the government to hire contractors. One such pressure is the assumption we debunked in *Bad Business* that contractor employees always cost less than government employees. Light wrote. Contract employees are less expensive only until overhead—or indirect costs such as supplies, equipment, materials, and other costs of doing business—enter the equation" (Ref. 197, pp. 1–5).

Effectively using the mass of contractors in "normal" situations is difficult enough, but when contracting is conducted during disasters, the possibilities of fraud, waste, and incompetence multiple. Recently, FEMA contracting in Puerto Rico demonstrates all of these problems with $3000 generators (that cost the contractors only $800), $666 sinks and subcontracting chains that featured each subcontractor doing no work but adding a few percent to the final cost so that the final FEMA expenditures bought much less than a single contract would have provided. Using the words "markups, middlemen and overhead" summarized the problem succinctly. In housing and repairs generally the money FEMA reimbursed each homeowner resulted in approximately half or less of what the normal price would have been. The bureaucracy around housing was, in fact, so complex that many applicants had seen no reimbursement until 5 months had passed after the Hurricane had passed. And of course, some of the contractors were shown to be major contributors to the political campaign. Apparently very little was learned from FEMA's contracting failures in New Orleans after Hurricane Katrina. If contracts are reviewed by credentialed (e.g., CPAs, when appropriate, etc.), adequately staffed FEMA employees most of this could have been avoided, assuming that political pressure was not applied up and down the process with favored contractors being selected time and time again. In fact, one of the "procedural" practices that freezes out honest local and even national contractors is the overall governmental practice of requiring contractors to have had a least a couple of profit producing years before being accepted as federal contractor. This informal "rule" has the effect of freezing out many potential contractors in favor of some larger, repeat contractors. Unfortunately, the aftermath of Hurricane Maria in Puerto Rico is an almost text book example of what can happen when contractors that are more skilled in political relationships than in technical skills are chosen. As this is written, Puerto Rico is still largely an area that has absorbed tens of billions of dollars but that has not "recovered" in any sense of the word.

Before closing this section on contracting, we will cite three case studies demonstrating some of the worst aspects of contracting

A brief case study: A truck driving contract for general conveyance after Hurricane Katrina

One of the coauthors was called by a relative regarding securing a truck conveyance contract during the time period just after hurricane Katrina, in New Orleans and in the adjacent state areas, particularly Mississippi. The relative was in the trucking industry and made his living by matching the users of trucking services and trucking companies in the Denver area. He was calling about a woman he knew owned about 10 trucks, 10 or more drivers, and normally serviced the areas of rural Louisiana, rarely working in the City of New Orleans or in adjacent areas in Mississippi. Her drivers were union members and were earning in the high $30s/h as I was told. I was not surprised by the call, as relatives or friends of friends in areas of disaster damage frequently find their way to FEMA or NDMS (in this instance) employees in understanding or making their way through the sometimes-difficult contracting systems. This in-law was told that while he could neither secure nor direct the contract he could point out where to enter the bid process.

One of the coauthors was already serving in New Orleans as a FEMA public health and medical liaison officer and was well aware of the impending need for many trucking contractors but was also all too aware of the small local and state emergency management agencies and their staffs, beset by problematic, if not often nearly nonexistent leadership. We were also aware of FEMA's difficulties in logistics management, financial accountability mechanisms, and overall lack of skilled upper management in the Washington, DC Headquarters, many of which had retired, resigned, or quit in the last few years and were frequently not replaced (some of these issues have been dealt with in other parts of this book). The deteriorating logistics situation was very clear with the lack of buses to ferry flood victims to and from the shelters, health care, and other needs. There was also an early lack of water, food, and related commodities distribution. Some FEMA logistics staff member was blamed and later fired, but none of this sped up the response.

After getting the contact information from the little truck company owner and clarifying that she was being dealt with as any other truck company would be dealt with, FEMA Headquarters was called and emailed alerting them to the availability of a trucking asset we knew was sorely needed, among many others. After a few tries with no response, the Regional Office Fiscal staff was called, and the information was faxed and emailed to them. The FEMA Regional Office fiscal staff was small, just a few staff members, but highly skilled and experienced. They outlined the newest process in place for contractors. The process was conveyed to the small trucking company's owner. Nothing happened after she put a bid in for many contracts. At that point someone advised the small trucking company to contact her Congressman and work with his office to get the bids recognized. That seemed to work when she received notice that one of her contract bids was accepted. Unfortunately, that was not the end of her contract efforts.

The bid was accepted by FEMA for something in the area of $10/h, quite a bit less than the over $30 some/h that was the normal union rate. Upon inspection of the accepted bid, and with some further research on the whole process it became clear that the original contract was bid and accepted at about $34/h, but was subcontracted at least 5 or more times, each time a dollar or two lower than the previous bid. We don't recall if depreciation, insurance, and fuel were built into the contract, but recall that there were some contracts adjustments for these issues. After some additional research it appeared that the subcontractors were not local concerns. Nonetheless, the small trucking company owner took the bid and was able to secure at least $10/h for her drivers, which they all needed just to avoid personal economic tragedies. The owner used her savings to make it all work.

A very brief case study: Outmoded communications gear in a response agency

One of the authors was serving as the Lead Regional Emergency Coordinator in a federal regional office (There are ten federal regional offices, in accord with the breakdown followed by most agencies that have local office across the country. The Regional office locations are usually in the largest city in the federal region. FEMA has long adopted this 10-regional strategy). The office was served by a local contractor from the building, a large structure approximately 2/3 of whose floors were populated by federal programs and federal staff who assured computer connectivity, network fixes, and related issues. The office had landlines that were at least 15 years old (finally replaced two or so years later), once featuring conference call options, call forwarding from a central location for each suboffice, usually run by an Administrative Assistant. The many phone options no longer worked, sound quality was relatively poor but the normal call screening worked well most of the time. A large national contract purchased the phones, the clunky, 7-year old portable computers but not the desktop computers, which were purchased by each agency. The contractor hired local IT staff, which were excellent, and managed to keep the old equipment working. Handheld phones were paid for by Headquarters and were fixed or repaired to the extent possible by the contractor, usually acting to help even though this may not have been called for under the large, national contractor's agreements. The large national contractor has extensive federal contracts, with the most, apparently, with the Department of Defense in a variety of capacities. There was no bidding of which we were aware of through at least three different US Presidents. The key element or takeaway here is that these were the basis of the communications capacities for a federal disaster response agency, an agency that might have been expected to have paid for and received the newest and best in communications gear and networks. This same, large national contractor provides similar services for a great many federal agencies.

This very brief case has been selected to present the Emergency Manager with a flavor of the huge impact powerful, national contractors have on federal agencies attempting to complete their missions. There is no implication that most federal contractors supply old, stodgy technology, receive their contracts continually, mostly in a no-bid manner, and are virtually unreachable by federal staff as they attempt to complete their missions, but of course too often they are. On the other hand, Emergency Managers need to be careful criticizing these contractors in public for even obvious failures. The large contractors lobby hard and have strong friends everywhere. In a more fundamental sense, they are part of a tradition trivializing federal traditions, with state and local governments often functioning more like Washington's administrative appendages than sovereign civic authorities. Statements like this are not pleasant to make but must be kept uppermost in the minds of Emergency Managers. In a somewhat cynical, but realistic vein, DiIulio summarizes his concerns and places them in a political context, in which big contractors are just beneath the surface. This following selection refers to the Obama Administration; there are similar examples from the two Bush, Trump, or other Administrations.

> ...congresspersons win reelection by fighting phony ideological wars with each other, lavishing debt-financing benefits on constituents, taking campaign cash from groups that get government grant or contract dollars, and using proxy administration to shroud government size and attenuate their accountability for its performance....Many for-profit businesses and nonprofit organizations related to government as narrowly self-interested factions. For example, drug companies virtually wrote certain Obamacare provisions, and in 2011 and 2012, organizations that won federal funds to implement Obamacare spent more than $100 million on lobbying (Ref. 191, p. 25 citing Ref.[199])

A brief discussion: KJ, an EPA subcontractor in environmental protection

KJ started as an Environmental Contractor approximately 20 years ago, when she was fresh out of school with her Master's degree in Environmental Sciences. She was making almost $90,000 per year. She had pension and medical benefits and enjoyed her job thoroughly. This year she is making about $80,000 annually, still enjoys her job but has no benefits in addition to her now stagnant salary. In the past she would have been able to transfer to another EPA contractor or even to the EPA as permanent federal staff. At present neither of those options is possible. KJ observes that she is not much different from a huge batch of contractors who vastly outnumber full-time federal employees working for the EPA.

Mitigation can be effective, but it is usually severely underfunded at all government levels

A major area in emergency management where money seems to have little relation to solving problems is in mitigation, an area in which FEMA funding has moved away from in recent years, for a variety of reasons, most having to do with "saving money." It is clear that states aren't spending enough on mitigation before a disaster occurs, though a major reason is that FEMA flow through funding for mitigation has mostly dried up, in favor of a push for state/local resilience. This is a good thing, but no substitute for at least partially funded mitigation strategies. But that should be no surprise, since all of emergency management is in more or less a

FIG. 2 Building in areas subject to floods and tsunamis leads to predictable damage.

scarcity mode that has only gotten worse in 2017 and 2018, despite the estimated $306 Billion in disaster-related damages in 2017, and an even larger total for 2018 (Fig. 2).

At one level this is not understandable, because it has been cited over and over again that for every single dollar spent in mitigation, approximately six dollars are saved in eventual disaster reimbursement.[200,201] If viewed primarily from the political level it is not surprising that mitigation funding is not necessarily a FEMA focus, because, as we have discussed, most political actors, elected or appointed, are oriented to upcoming election cycles, whether they are one, two, three, or 4 years away, are focused on the here and now. Unfortunately, mitigation savings are experienced during future disasters, where money and perhaps even lives are saved, but nonetheless in a somewhat distant, and certainly mostly unforeseeable future. This is more than a theoretical perspective we are summarizing for Emergency Managers, it is where they live and work.

There is a scarcity of information on overall disaster spending at federal, state, and local levels

If it were not already difficult to allocate funding for mitigation efforts, the sad fact is that in most instances the amounts of resources spent on natural disaster assistance, especially at the state and local levels, is too often an unknown quantity. Neither FEMA nor any other federal agency keeps this anything approaching comprehensive information on overall disaster spending. Most states or local areas do not have the procedures or the resources to comprehensively track natural disaster spending. The episodic nature of disasters generally means that policymakers struggle to keep their focus on tracking costs. Further, data collection can be especially hard because there are often numerous state agencies participating in the various aspects of disaster assistance and they often report what information they provide differently. Also, state spending is highly variable because of diversities in state resources, likelihood of natural disasters, and differences in the political support levels for emergency management.

Helping communities prepare for, respond to, and recover from events such as storms, earthquakes, and wildfires—involves a surprising array of players from the public, for-profit, and nonprofit sectors, almost none of which have anything to gain at present, from coordinating data and information gathering and storage during and directly after disasters. Central to those efforts is a complex intergovernmental partnership that is already under stress from the increasing frequency and severity of losses and from the all too prevalent budget constraint at all levels of government and not-for-profit sectors.[200]

If overall natural disaster spending is not well tracked and accounted for, it should be no surprise that mitigation spending is also not comprehensives tracked despite the nearly $140 Billion Congress provided in one-time funding for 2017's historic hurricanes and wildfires.[202] This lack of information occurs despite the nation's strong progress in hazard mitigation science. In the last 15 or more years this has included strong progress in plotting hazards spatially, targeting the areas of greatest risk, and identifying and implementing appropriate risk reduction strategies often using imaging, sensing technology, including satellite imagery and aircraft-based systems (such as radar, LIDASR, and FLIR systems), including even the rapidly expanding use of sophisticated drones (Ref. 2, p. 73). Of course, policymakers are looking for effective ways to control costs by investing in mitigation activities that will reduce the risk to lives and property before a disaster happens, or more usually, before the next disaster happens, though actually seeing the present-day investment is not as easy as one may expect, given the political realities addressed before.

Although a relatively bleak picture of the lack of adequate mitigation spending at state and federal levels and the accompanying lack of comprehensive spending not just on mitigation, but on overall disasters spending, state to state mutual aid is a bright spot where relatively small, pointed expenditures can be extremely effective. For example, a week after Hurricane Harvey made landfall in southeastern Texas in 2017, at least 21 states had sent a variety of emergency response teams and equipment to help. And each dispatch of assets is based on previously crafted agreements, often backed up by mutual trainings and exercises. The compact began as a regional effort among southeastern states and was formalized in federal law in 1996; it now includes all 50 states, the District of Columbia, Puerto Rico, and the U.S. Virgin Islands providing at least $800 million in personnel and equipment to help Alabama, Florida, Louisiana, Mississippi, and Texas to recover from widespread damage.[248] Needless to say, effective participating was not the norm in Puerto Rico in 2017.

FEMA has too many recurrent weaknesses to be a consistent good steward of its money

Despite its importance in the costly and every growing sector of naturally occurring disasters, FEMA is subject to the frequently inexperienced, politically appointed upper levels of management that plague most other federal agencies. And with its relatively small crew of full-time federal employees and its high percentages often underpaid and under skilled temporary disaster assistance employees and contractors, FEMA is subject to the same administrative complexities that afflict most other overly outsourced federal agencies, as described before. FEMA's disastrous response to Hurricane Katrina in 2005 and its role in the post-Katrina human, physical, and financial recovery process has been harshly criticized as has its equally disastrous response to Hurricane Michael in Puerto Rico in 2017. But its many critics have failed to notice that the FEMA that failed so spectacularly in the Gulf Coast had in the preceding decade or so become home to a downsized, inexperienced, and overloaded federal workforce and with expanded missions, including terrorism response and preparedness. Then, for the better part of a decade FEMA received positive Congressional and Presidential Administration support, for the most part, and was allowed to hire additional staff and received higher levels of fiscal support which enabled the agency to conduct a generally effective response to the huge, cold weather Storm Sandy in 2013. During the Sandy response FEMA had about 4800 full-time employees and up to 6700 reservists, many of whom could benefit by higher levels of training, including its variable contract workforce. But FEMA is always hard to predict, and always subject to the next election, so, despite this recent progress and response success, a new Administration came on the scene and the Puerto Rican tragedy occurred, soon to be followed by good responses in the Hurricanes of 2018 (Ref. 191, pp. 59–61).

Emergency Managers are required to work with FEMA at a variety of levels and through a variety of mechanisms. As they do so, it is important to have conducted at least a quick review on the status of FEMA's preparedness and abilities to adequately control the money it both spends and monitors. Having a few contacts at the local, FEMA Regional Office will remain one of the most important set of contacts and colleagues across the disaster field. Spotting FEMA's trends, such as its recent moving away from strong support for mitigation, underfunding its own fiscal review staff, whether or not FEMA permanent staffing is on an "Upswing or downswing" and whether its missions continue to outpace its resources are areas of interest to any Emergency Manager in the field. Effective decision making in emergency management will continue to depend on much more than good disaster response and recovery talents.

The full-time federal workforce

Now that we've discussed ongoing FEMA weaknesses, that by and large are not the faults of the permanent federal staff, it is time to address that staff itself. Following we'll also deal with agency cultures and the motivation of federal staff. The stereotype of FEMA's permanent federal staff, in fact any permanent, full-time federal staff has been presented by Charles Murray, a consistent critic of the federal workforce and the overall federal government who nonetheless makes some insightful observations. But it is his negative views, for the most part, that are very reflecting the view held by many Americans as well as by many members of Congress, regardless of political party.

> Murray characterizes federal bureaucrats as quite a sorry lot: ill-trained, "second-rate people" who, to "fill the many quotas" that govern the federal personnel hiring process, are "routinely promoted a few steps beyond their levels of competence;" they work in "often drab and poorly maintained" offices (nothing like the private sector's "attractive, open spaces with bright colors, lots of light" and room for "creative interactions among staff"), and yet get paid "more than they could" command "in the private sector...Murray concludes with a virtual "counsel of despair": reelection-seeking incumbents, special interests, and bureaucrats "with prospects of going through the revolving door" create a "mutually reinforcing network of forces that will keep...reforms from coming to pass... (Ref. 191, pp. 129–29, 139)

DiIulio responds wisely to Murray:

> Behind ... reports about federal programs "duplication" and "overlap" and waste are not incompetent bureaucrats exercising renegade discretion, but lousy lawmakers and diverse sets of potentially entrenched proxies, each of which gets to wet its beak and get its funding without regard to the most rudimentary principles of cost-effective public administration. Nobody who wanted to translate good laws into good administrative actions, and who was the least bit sane, would design... from scratch and...having ever fewer people handling more public money, doing ever more (and ever more complicated) tasks, and managing ever expanding disasters that the federal government has experienced as well as a guarantee of more—and worse—disasters to come, unless and until we not only hire more full-time federal workers but train, motivate, performance mag, and reward them as if our public well-being depended on it (Ref. 191, pp. 141–42).

As discussed, Murray has stated what appears to be the conventional wisdom about federal employees, but many of his beliefs do not appear based on fact. For example, there is not constant cross-over between federal employees and the private sector, especially at the upper level. The average Senior Executive Service member, those who can be expected most likely to "jump ship" for higher private sector jobs have been federal employees for at least 20 years, and don't even reach their highest grade until they are well into their 50s, with salaries in the $160,000–180,000 range, hardly what top managers in the private sector would make for managing program with tens or hundreds of millions of dollars or more.[203]

Retirement continues to be the major reason that most SES leave their jobs, at 61%. Work environment issues continue to be the highest contributing factors to SES leaving their jobs before retirement, with 42% leaving because of the "political environment" with "senior leadership" at 40% another leading cause of leaving before retirement age.[204] The authors have managed, and worked in federal, state, and local emergency response agencies, have contracted, worked on grants and have viewed the Senior Executive Service as self-styled members of a club whose members protect each other, keep their own counsel, but do not necessarily confide in many members of the overall federal workforce. Their jobs can be rewarding at many levels, but the difficulties implied by interfacing political appointees who have ultimate program authority and federal employees who know and implement the programs can be challenging. When SES members leave before retirement age it is usually these issue that drive them out, and not the attraction of higher salaries in the private sector.

Each federal agency has a distinct culture, just as each individual private nonprofit or for-profit business has a distinct ways of hiring, firing, and otherwise managing personnel in doing what they do. It may not always be readily apparent why some agencies have cultures that are associated with high-quality staff and high-quality outcomes, such as the Forest Service, The Army Corps of Engineers, and the Bureau of Prisons. These agencies have been referred to as having elite reputations that would be more difficult for Congress or Presidential Administrations to tamper with.[51] In fact, FEMA has long depended on "The Green uniformed" Forestry Administrative and Fiscal staff assignees to help FEMA properly administer a disaster in the long months of recovery, and occasionally even in response mode, if they have been requested early enough in the course of the disaster. The US Army Corps of Engineers has also had a long history of being one of the federal agencies that FEMA heavily depends on for assessing and addressing various physical aspects on mission assignments or on their own agency statutory authorities. Wilson stresses that the management of federal agencies is powerfully constrained by limitations regarding hiring, firing, fairness, purchasing, and generally keeping federal staff relatively free of political constraints on their doing their jobs. Surprisingly his massive study did not spend much time on the outsourcing dilemma that severely hinders governmental efficiency in so many ways obvious to those who have spent the time to see the many examples. Among his many conclusions, Wilson found that things to be "not so bad":

> After reading this long recitation of the constraints under which government works, you may think that I am pessimistic....Not at all. Every democratic government has these problems, and our(s)... does better than most others in attending to these defects and trying to fix a few of the worst ones. In fact, by the standards of most democratic governments, many agencies are friendlier and more cooperative than their counterparts abroad. Our constitutional system so fragments authority and encourages intervention that it produces two opposing bureaucratic effects: citizen-serving agencies that are friendlier and more responsive and citizen-regulating agencies that are more rigid and adversarial (Ref. 51, p. xv).

FEMA is a service agency, and those saying its field, headquarters, or regional staffs are generally unfriendly and unhelpful to citizens are probably not being truthful. It is likely that elite agencies, and even more mundane agencies draw staff that are sincerely attracted by their missions, it is not likely they are attracted by the high salaries, which often just don't exist. (There is an exception for especially skilled, highly necessary staff, for example, staff with strong vaccine-development credentials, who can make in the area of $300,000, well over the federal pay scale. Unfortunately, sometimes these salaries are given out on what appears to be almost arbitrary grounds.) On the other pole are agencies like the Internal Revenue Service that are more rigid, and fortunately for Emergency Managers, the IRS rigid tendencies are not ones that need to be dealt with in the field. The irony is that FEMA has both poles within itself. Its service, disaster response and recovery staff and actions are friendly, if not always highly effective and timely. But its regulatory, rules-oriented staff, especially those dealing with reimbursement under such responsibilities as the flood insurance program or even public assistance titles can be rigid, often maddeningly slow in response, and too often cruel in their seeming adherence to rules that sometimes defy common sense. These situations are exacerbated by the complexity of FEMA's state and local agency missions, various other outsourcing, the shortage of full-time FEMA staff, and the often inexperienced disaster reservist staff. If this all weren't difficult enough, many FEMA staff decisions are public and have exposure to strong political considerations regarding declarations of disaster, the percentages of required state/local cost share, the occasional public relations disasters and similar issues.

Emergency Managers at all levels must deal with both FEMAs, the friendly one as well as the more rigid one, even occasionally getting those higher in their chains to seek Congressional assistance in those instances where justice or unwise decision making is at issue. It should be clear that the "rules" and the large expenditures are more often recovery issues as opposed to the early response issues, which are more enjoyable to deal with, more satisfying usually, and of much shorter duration. Those who study disasters remind us of the long history of squabble between the responders and the recovery and mitigation folks Scott Gabriel Knowles.[205] Based on all of this is very smart to be extra nice to FEMA's fiscal staff as well as the fiscal staff of the various ESFs that the Emergency Manager encounters in the duration of a disaster. Of course, acting poorly to anyone under any circumstances is a request for difficulties at many levels.

Trying to define what motivates a federal bureaucrat, those that work for FEMA, The US Army Corp, the Forestry Service, and the Environmental Protection Agency in particular, because of their special importance to Emergency Managers is another difficult task, and one that we have addressed in a number of perspectives all throughout much of this book. Alissa Martino Golden[52] (pp. 171, 72, 8–26) chose to address this topic after studying the Reagan Administration, a period in which the President used his authorities to change the size of as well as the way many federal agencies and federal employees both considered and accomplished their jobs. There is no intent to summarize how the various agencies and individuals reacted, instead, the Emergency Managers reading this will be exposed to Golden's general findings, which appear to be useful in almost any time period. Her findings are listed as follows:

- The Office of the President has the authority to select agency leadership at the G.S. 13-15 and SES/Schedule C (Political Appointee) level to conform to his (or her) policy direction, regardless of the preferences of the federal career employees. The President's general guidance will tend to set the tone for agency policy change and implementation.
- Despite this authority, specific policies can still be pulled in directions in which the professional training and values of federal employees consider both legal and wise. It is at this level that policies that federal employees feel are not in the best interests of the country can be moderated in many instances.
- It would be unrealistic to believe that all federal employee pushback against an Administration's policies is based on altruistic, patriotic, and professional motives. As employees, federal bureaucrats resent losing benefits, grades, nice offices, and other perquisites minimized by a new Administration.
- Agency "Esprit de corps" is a powerful tool that an elite agency, as well as other agencies, can employ to moderate policies for both altruistic as well as more self-interested motives. Golden mentioned

that under the very popular President, Ronald Reagan, the EPA lost staff and was stymied in enforcing rules and giving grants in ways that it felt (but not the Administration) were not appropriate. On the other hand, we recall the old FEMA cliché, "When everyone else is running from a disaster, we run to it," which is a continuing ideal to which many FEMA response staff will probably always aspire. And of course, even when the bureaucrat acts in his or best professional judgment, this may not be a winning formula when a new Administration has a widely different perspective.

And finally, to complete this extended discussion of federal outsourcing and cost efficiency, it will be helpful to review some of Michael Lewis's comments about federal employees as he reviews recent years of sometimes deteriorating government processes. Why does someone go to work inside this little box—or any little box—inside the federal government? There's always an answer to this question. And it's obviously important. Why a person does what he does has a big effect on how he does it. Using the example of the National Weather Service that provides virtually all of the data that private weather companies (e.g., The Weather Channel, AccuWeather) use, but is forbidden by law from advertising the value of its services—and if it even hinted at doing so, the private companies could apply pressure on the National Weather Service in a variety of ways. So, when some of the private industry competitors made unsubstantiated or even false claims about their accuracy in weather predictions as opposed to the National Weather Service, the meteorologists at the National Weather Service had no real ability or even inclination to defending themselves against false charges. According to Lewis, "They (The National Weather Service's federal employees) never claim credit. They always do these intensely self-critical how-can-we -do-better inquiries. It's a public safety mentality, they do what they do because they really sincerely and since they were eight years old, they love the science and the service, not because they care at all about credit or glory." That was the sad truth—public servants couldn't or wouldn't defend themselves and few outside the US government had a deep interest in sticking up for them. Barry Myers of AccuWeather said to the McKinsey consultant firm in a study of the National Weather Service "…the National Weather Service does not have the final say on warnings…. The customer and the private sector should be able to sort that out.[206] The government should get out of the forecasting business'. Of course he didn't say the government's National Weather Service should stop generating all of the data that the private sector used mostly without attribution.

7

Does the National Incident Management System (NIMS) really work for major event management?

It's like you're bilingual...you can speak the language, but you just can't read it. –State Trooper, commenting on ICS

Because of the September 11, 2001, terrorist attacks, the US Homeland Security Act of 2002 mandated the creation of the National Incident Management System (NIMS) to be the standard method for managing emergency response operations at all levels of government regardless of incident type, size, or complexity. Millions of responders, including Emergency Managers have taken classes. Is the NIMS much more than the ICS? In reading the NIMS document, much of the content is taken from ICS 100–400 course material. The ICS has the potential to work well as an incident management tool but there are consistent problems. Lack of knowledge by users and events that are greater in scope and complexity that the tool was designed for.

191

One of the authors participated in providing ICS 300 and 400 training to members of the committee that built the first NIMS guidance document. What was telling is the class was given 2 years after the first NIMS document was published. How could guidance be developed by people who were not subject matter experts? With the bounty of subject matter expertise in this country, I think we could have done better. Was an exhaustive review of incident management systems in use around the globe undertaken? How about ISO 22320 from the International Organization for Standardization that serves as an international standard for incident management? (Fig. 1).

To ensure that ICS is used across the country as extensively as possible, NIMS/ICS implementation requirements starting in the year 2005 gave jurisdictions 2 years to comply with the full array of NIMS/ICS implementation standards. NIMS/ICS compliance was made a precondition for any agency or organization to receive homeland security preparedness funding. What agency is going to give up being federal grant eligible? Seems like a great idea. However, there never was an actual withholding funds from jurisdictions that did not comply with the NIMS/ICS mandate. It proved too difficult and even could be considered counterproductive to enhanced preparedness.

Withholding funds would have removed resources that those entities needed to improve emergency response systems, and that would undoubtedly have caused political reaction by local, state, and federal policyholders representing those jurisdictions. As a result, states and substate jurisdictions, when applying for homeland security grants, have only been asked to self-certify, with minimal documentation, that they are NIMS compliant. Thousands of organizations and jurisdictions have documents signed by their chief executive saying they are NIMS/ICS compliant.

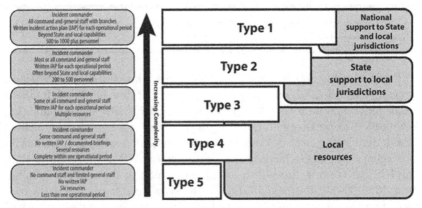

FIG. 1   National Incident Management System event typing system. This gives jurisdictions across the country the ability to consistently label events.

If in fact NIMS/ICS was the answer those trying to manage major events were looking for, we should see have seen improvement with all these organizations now NIMS/ICS compliant. This doesn't seem to be the case. Compliant does not equate with competent and this is where it's hard to make an argument that NIMS/ICS are working in major event management.

When Hurricane Katrina struck the Gulf Coast in August of 2005, the NIMS/ICS mandate was about a year and a half old. Some will say that this had not been sufficient time to integrate NIMS/ICS fully into emergency response procedures prior to the storm. That may be true. What's also true is that one of the authors was in Baton Rouge months before Hurricane Katrina teaching an ICS 300 and 400 class prior to hurricane season. A senior officer with one of the response organizations met the author at the venue the first morning to unlock the building. When the author asked them if they would be in class that week, the senior official said, no they had had all the training on the topic they needed. It's fair to say the transportation sector struggled in its emergency response efforts, especially with respect to inadequate prepositioning of transportation assets for evacuation needs and incomplete assistance to special needs populations.

If the challenges of command and control, resource management, and communication during Hurricane Katrina were due to the relative newness of NIMS/ICS, time passed should reveal improvement. The following findings were taken from FEMA's After Action Report (AAR) produced after the October 1, 2017 Mandalay Bay shooting. Thirteen years have passed since NIMS/ICS was mandated.

- Prior to the incident, command was not unified at the Route 91 Harvest Festival. LVMPD and Community Ambulance were both operating in the venue as planned; however, CCFD was not integrated into the special event plans or operations.
- The first CCFD unit on scene, Engine 11 (E11), did not establish command.
- Communication difficulties with key medical providers complicated response efforts.
- The LVMPD Communications supervisors and managers on duty on the evening of the incident could have benefited from additional training and support to effectively manage an incident of this magnitude. Additionally, some Fire Alarm Office dispatchers mistakenly communicated with the Operations Section and branches directly instead of Incident Command.
- Some fire department responders were confused by the use of cardinal directions, rather than local landmarks, to define divisions under branches.

- The fire department EMS branch director role was assigned twice. However, both branch directors were operationally ineffective, as EMS was being handled in the North and South branches. The branch directors failed to notify Incident Command or Operations of their operational ineffectiveness.
- The fire department's North and South staging areas were not managed effectively and had difficulty coordinating with one another.
- The large number of ambulances provided by multiple private transport companies facilitated transport of victims but also made command and control difficult.
- Fire department span of control issues hindered information sharing, which in turn resulted in challenges for Rescue Task Force teams in locating and treating patients.
- In an attempt to release outside jurisdictional fire department resources, the Operations Chief advised South Branch to release HFD units to clear and return to service. Some of these crews were actively conducting floor sweeps at Mandalay Bay when this order was issued. This decision left the remaining law enforcement component of the Rescue Task Force teams understaffed with several hundred more rooms to search.
- There were communication and coordination shortfalls among officers related to clearing the venue due to the lack of a forward command.
- The CCFD Incident Commander was unable to leave his command vehicle because he needed to monitor radio traffic on multiple channels simultaneously. This confinement to his vehicle created initial challenges in establishing Unified Command with LVMPD.
- The fire department Operations Section Chief and branch directors would have benefitted from representatives from outside agencies assisting them.
- As the scope of the incident expanded on the Las Vegas Strip, there became a need to expand Incident Command System roles and assignments.
- Individual responders circumvented command and requested resources to the scene without Incident Command knowledge or approval.
- CCFD's use of multiple tactical radio channels facilitated the setup and expansion of the Incident Command System and command framework. However, expansion of the communications plan led to confusion among some crews on scene and inbound, as well as among Fire Alarm Office dispatchers.
- Radio signal issues in certain areas of the Mandalay Bay prevented first responders from transmitting or receiving crucial information in some cases.

Findings such as these are not hard to find in every AAR produced after a major event, often with terms like "lessons learned." Either we are not learning the lessons, not taking the right class or we've been given the wrong answers.

It's a little harder to interpret FEMA's AARs of its own efforts as they tend to avoid NIMS/ICS verbiage and instead use terms such as whole community, survivor-centric, and other focus areas developed by current leadership. Not exactly an apple to apples approach other's performance is measured against.

Surely the agency (FEMA) responsible for giving us NIMS/ICS would be competent in 2012, 7 years after mandating everyone adopt NMIS/ICS, right? The following findings are from the Hurricane Sandy report.

- Area for Improvement: Integrating Federal senior leader coordination and communications into response and recovery operations.
- Area for Improvement: Coordinating Emergency Support Functions (ESFs) and Recovery Support Functions (RSFs) to support disaster response and recovery.
- Area for Improvement: Implementing incident management structures.
- Area for Improvement: Coordinating among states, localities, and tribes.

Of interest to the authors, in the earlier AAR it was recognized that FEMA needed to improve having and being able to deploy a qualified workforce. One of the major challenges noted in the 2017 hurricane season was the lack of a trained workforce. Lesson's learned aren't always learned, are they?

This is also troubling since the latest revision of third edition of NIMS was published in October of 2017 and spoke of the NIMS qualification, certification, and credentialing process using a performance-based approach. The intent was to use nationally standardized criteria and minimum qualifications for positions provide a consistent baseline for qualifying and credentialing the incident workforce.

Along with the job title and position qualifications, the position task book (PTB) is a basic tool that underpins the NIMS performance-based qualification process. PTBs describe the minimum competencies, behaviors, and tasks necessary to be qualified for a position. PTBs provide the basis for a qualification, certification, and credentialing process that is standard nationwide. Problem is that FEMA recommends minimum qualifications, but it is the Agency Having Jurisdiction (AHJ) across the Nation that establish, communicate, and administer the qualification and credentialing process for individuals seeking qualification for positions under that AHJ's purview. That may work for an individual agency, but who is doing anything competencies of personnel from multiple agencies and disciplines and levels of government that need to come together to manage a major event?

In 2017 a study was conducted (Coordination in Crises: Implementation of the National Incident Management System by Surface Transportation Agencies, https://www.hsaj.org/articles/13773). There are some interesting findings that support the argument that NIMS/ICS may not be the answer to the question of ideal major incident management.

Post 9-11, "terrorism" was the big driver but over time guidance documents switched to "all hazards" preparedness over terrorism-specific preparedness. References to the NIMS/ICS began to emerge and transportation agencies started taking their ICS classes as recommended by the USDOT and TSA. These courses included IS/ICS 100, 200, 300, 400, 700, and 800. FEMA even developed the publication *The National Incident Management System – A Workbook for State Department of Transportation Frontline Workers* as a replacement for ICS 100 training, and the Transportation Research Board funded the creation of ICS training for field level transportation staff through the Mineta Transportation Institute in conjunction with the California Department of Transportation (Caltrans).

A lot of work trying to help one agency that plays a huge role in major disasters institutionalize NIMS/ICS. However, transportation-specific NIMS compliance standards were not available, and where guidance has been issued, it is not consistent. For example, the USDOT and Transportation Research Board (TRB) have issued guidance documents with NIMS compliance recommendations. The recommendations vary and, in both documents, transportation agencies are advised to consult with their states' emergency management agencies for specific NIMS compliance requirements for the sector because the federal government has delegated creation of discipline-based standards to state governments. Not the clearest path to becoming NIMS/ICS compliant, let alone competent.

State and municipal transportation agency representatives across the country were asked what are the factors that have helped or hindered them in adopting and using NIMS/ICS some 12 years after it was mandated.

Significant factors that helped were as follows:

- Relationships/collaboration with emergency response/management agencies
- Executive support
- Perceived Risk from Previous Incident

Minimally influential factors:

- Grants that support or require NIMS/ICS implementation
- After-Action Reviews Supporting NIMS/ICS Utilization
- Laws, Regulations, and Policies Requiring NIMS/ICS Utilization
- Free Online FEMA Training

Significant factors which impeded NIMS/ICS Implementation were as follows:

• Limited grant funding or budget allocation to NIMS/ICS activities.
• NIMS not a priority or viewed as mission critical by agency administration.

Minimally influential factors:

• Low perceived risk from infrequent events
• FEMA courses—accessibility and content
• Lack of or unclear NIMS/ICS compliance standards for agency

It's clear that a good deal of effort went into trying to help transportation agencies buy into the NIMS/ICS program but there were and are significant challenges. Now think about all the various organizations who didn't put in near the effort that the transportation sector did. How well do you really think NIMS/ICS is being institutionalized across the country?

Is it time to face the ugly truth and ask the question, if NIMS/ICS hasn't solved issues of command and control, communication, and resource management in major events, what do we do next? Throw out NIMS/ICS? Look to other countries for management solutions? Mandate more? The authors suggest that we've invested too heavily in NIMS/ICS to choose another route without someone losing face. There's no argument that topics under NIMS/ICS are important and that the ICS is a great tool for management of many events but based upon performance since NIMS/ICS was mandated has not improved significantly. Even use of the relatively simple Unified Command feature of ICS is not always used when appropriate. To expect the more complex organizational choices offered in the ICS 300 and 400 courses seem like an impossible dream. For ICS use, there appears to be a ceiling to its application (Fig. 2).

It was never going to be as easy as a Congressional mandate for NIMS/ICS, to immediately bring tens of thousands of government entities at the state, county, municipality, and district level together? To see the benefit of NIMS/ICS or a standardized response during major events it must be used across levels of government, Native American tribes and jurisdictions, be accepted by diverse professions, take root in hundreds of thousands of individual agencies and organizations, and spread through the public, private, and nonprofit sectors. At the organizational level, there is not one person assigned as the NIMS/ICS-Liaison. Successful implementation requires institutional adaption by all agency personnel at the operating level. A tall order.

Despite the known drawbacks and lack of enforcement of compliance let alone competence, FEMA continues to foster NIMS/ICS implementation as the answer. This has included revision of guidance documents

FIG. 2   Complex events require complex event management skills sets.

to all levels of government, as well as to private industry and nonprofit organizations. FEMA continues to create NIMS/ICS training resources. Nothing wrong with training and it's clear that use of the ICS is lacking at smaller events, so it's obviously needed. This raises a whole other set of issues with current ICS training missing the mark.

By putting out training for specific disciplines, including transportation, health care, hospitals, EOCs, higher education, schools, public works, public health, and volunteer organizations feeds the problem. The ICS is discipline blind, so rather than ensure a solid understanding of the core principles and allow students to use the ICS tool as they see fit based upon their organizational needs, classes are making it more confusing. Understanding the differences among professions that participate in emergency response really isn't the problem, it's how do we work in a coordinated fashion? In the experience of the authors, it is often relationships, either preexisting or developed during the heat of battle.

The Emergency Manager is put at significant disadvantage when it comes to major incident support and management, and the challenges are not solved though NIMS/ICS. Resources available through government channels are dwarfed by those available in the private sector. A wise Emergency Manager knows logistics will be an issue during the response and recover efforts would be miles ahead by fostering a relationship with their local FedEx or UPS operations manager. Suggesting these companies couldn't make the life of the Emergency Manager easier unless they were NIMS/ICS compliant or course completion certificates in files back in their office are ludicrous.

Emergency Managers have semiregular contact with organizations tasked with emergency response duties and/or emergency management functions. The field of private and volunteer organizations with support potential during a major event but are outside of emergency management core missions can be massive.

In many cases during large events, emergent behavior emerges regularly that can take the form of *ad hoc* developments, either of groups or plans that are not congruent with the assumed "prepackaged" nature of ICS. Unsolicited volunteers and self-dispatched resources are ready examples of emergent behavior immediately following disaster impact.

NIMS/ICS compliance won't solve the challenges of how to incorporate them in the emergency management effort. The Emergency Manager needs to focus on objectives and who brings the solution to meet the need most effectively. When Emergency Managers fail or are slow to make these connections, you'll see private and volunteer organizations deploying outside the "official" response.

For example, the Cajon Navy has three core missions:

- Rescue—Our first plan of action in any disaster is to complete our rescue missions and save lives. This means going out on the front lines while disaster is still striking and saving all the lives we can.
- Relieve—Our second plan of action is to bring relief to those who have been affected by tragedy. That means making sure everyone has clean water, fresh clothing, hot meals, and everything they need to feel comfortable and safe.
- Rebuild—Our third plan of action is to help people rebuild their lives after a disaster. From demolition, cleanup, organizing and distribution of supplies, we want to make sure every family has a place to call home again.

The Cajon Navy is all volunteer and relies on donations. How will it look to the public if the paid Emergency Manager is too focused on building a common operating picture to immediately deploy rescue resources available to them through mutual aid agreements? Worse yet they make a disparaging comment of how the Cajon Navy is free-lancing and not working with the local emergency management organization? More important than how the Cajon Navy's operation upsets the local Emergency Manager is the negative impact on those who need lifesaving and sustaining services. This gets back to the skill of making good decisions with less than optimal information and less on being skilled at following the official deployment process.

Even if the threat of withholding federal grant funds worked and those entities relying on these funds were complying with NIMS/ICS guidance, we're missing the boat. Private industry, NGOs, and volunteers are not relying on this funding. Does that mean they should be excluded from the

Emergency Manager's playlist even though we see them playing a large role in major events? Both as impacted organizations and resources for the community or wise Emergency Manager.

Many of the private, NGO, and volunteer organizations have little if any budget for emergency management programs. Additionally, these organizations may perceive NIMS/ICS as overly prescriptive and rigid and therefore unsuitable for their operations. This also supports the argument that ICS has gone merit badge. Let's say a company offers kitchen supplies to new and existing restaurants. In a major event they could prove valuable in helping establish mass feeding stations at shelters or available buildings. What ICS class should they take? An online class which gives them little competency? Take any class that is offered close to their office? Lobby FEMA to develop a kitchen supply company ICS course?

Without some relaxing of NIMS/ICS guidance or ability to flex and customize NIMS/ICS in ways relevant to each individual organization's needs, structure, and culture they will never be incentivized to participate. The Emergency Manager may be a subject matter expert in NIMS/ICS, but if they don't the same level of expertise in decision making, collaboration, and relationship building, operations will suffer.

Fire departments were first to widely adopt the use of the ICS and have gotten good at using the ICS to the extent they use in daily fire operations. Does a tool like the ICS, born out of California wildfires, really lend itself to be an "all-hazards" solution? Without regular use of the ICS, there will be issues of consistency of use. Infrequent utilization of NIMS/ICS is another obstacle to full implementation, particularly by most organizations involved in major event management, response, and recovery efforts. Asking organization to use an unfamiliar system during the worst event of the lives and expecting good results seems beyond wishful thinking.

Is it that important to be competent in the NIMS/ICS?

Organizations and agencies at all levels are required to demonstrate compliance with NIMS in order to receive federal funding for emergency management programs. Currently to demonstrate compliance, these agencies must show that they have adopted ICS as their chief emergency response system and that their personnel are adequately trained and can function in the system during a response. The question becomes, so what happens when performance indicates inadequate training and incident management personnel can't function?

Emergency management law in the US is found at the federal, state, and local levels. Keep in mind that all the guidance and federal directives post-9/11 were designed to fix the problems seen during 9-11, for example, issues with command and control, resource management, and communications. Coupled with the National Fire Protection Association (NFPA) and we now have new standards that apply to all Emergency Managers. Use of ICS is *required* by both NFPA 1600 and NIMS.

In the context of the Emergency Manager, negligence could arise from the failure to perform (or unreasonably bad performance of) specific governmental duties. The unit of government may incur liability from failure to properly train or supervise emergency management workers. Other frequent sources of liability include failure to perform the duties that are generally accepted as being part of emergency management's responsibilities. Expecting Emergency Managers to be competent in NIMS/ICS is not a stretch.

One of the basic duties of any Emergency Manager is to develop an Emergency Operations Plan (EOP). If possessing such a plan improves response, it follows that the court might entertain the question whether lack of a plan would be the basis for a lawsuit. All states have statutory requirements to prepare an EOP.

Failing to prepare a mandated plan will result in liability. Another situation that may result in liability is where a plan exists but may have been negligently drafted. Boiler plate plans are easy to create and will have checked all the boxes found in state and federal grant guidelines. The problem is that these boiler plate plans often include topics, services, and capacity not found within the jurisdiction. Saying you can evacuate your citizens at risk of flash flood may earn you grant dollars. Not being able to implement that plan, resource, and accomplish it, resulting in loss of life, may earn the Emergency Manager charges in court.

Other sources of liability may be found in executive level decision making (poor choices, poor planning, bad emergency response), or even an Incident Commander's lack of wisdom in decision making.

Additional programs create standards of performance for Emergency Managers. These include the Emergency Management Accreditation Program (EMAP). Although EMAP accreditation is voluntary, the fact that it is endorsed by such a wide variety of authorities means it is well on its way to becoming the *de facto* standard for emergency management in the United States. The more programs become accredited under the standard, the more likely a Court will be to hold all emergency management to the norm.

Most counties participate in the Performance Partnership Agreement (PPA) and the Cooperative Agreement (CA) with FEMA through their states. These documents explain the stipulations that must be accomplished before the federal government, through the state emergency management agency will release emergency management funds to local units of government. Each SEMA cooperates with local Emergency Managers to condense these outputs into detailed assignments, which are called "Compliance Requirements." These standards act as a form of private law, binding states and localities that receive funds to obey federal mandates such as NIMS/ ICS standards. Some counties do not receive federal funding through this funding mechanism, but most counties do which again creates an industry benchmark that complements NFPA 1600. A court may hold this to be a standard of care, creating potential legal liability for those units of government with noncompliant local emergency management offices.

It is important to be competent, not just compliant, when it comes to NIMS/ICS as liability occurs when we fail to use them properly and people suffer consequently.

Is our ICS training model broken?

The authors believe the ICS is best described as a tool. Tools are selected based upon their practicality for helping the user carry out a task safely and successfully. To accomplish this training must ensure the operator is competent to know when use of the tool is applicable, safety considerations, and trouble-shooting.

Considering millions of people have taken ICS courses over the past decade, it's disturbing to still see the lack of proper implementation and utilization when appropriate. While the ICS undoubtedly can help immensely at emergency scenes, it may not be the answer for Emergency Managers. Is it the ICS tool that's not working or is it the education and training we're giving people that is the issue?

If you accept for the moment that the ICS is a tool, training should be like that offered with other tools. A few suggestions are offered as follows:

- Our courses on ICS talk about scene safety a great deal, but not safety as it relates to the proper, or more important, improper use of the ICS itself. Flexibility of the ICS is a great benefit, but it also means selecting the right components must be used for the ICS to function without hazards.
- We need a well refined ICS systems check in place to make sure that the components and organizational structures we've selected are best for a given situation. Flight dispatchers create the flight plans for the Captain to review and enter into the flight computer. Checks and balances are key to safety every time the plane leaves the ground. The dispatcher and pilot own the process together. Let's look at one airline, Delta for example. On average the checks and balance of captain and dispatcher safely completes from 4800 to 5200 flights daily. There is nothing simple about the task of moving thousands of people so why wouldn't Emergency Managers want to learn more about how their system works to ensure life safety of similar numbers? An argument might be that that is the job of flight dispatchers and airplane captains.

Is the success of incident management at a major event any less important? This points out another problem we have with NIMS/ICS being implemented and used properly. We don't have enough people whose job is incident management, we're relying on part-time help.

- We need to provide more case studies in not just ICS success stories but emphasize the dangers that occur when the wrong components or organizational options are chosen. Lack of proper use of the ICS can endanger not just the people working but result in mission failure.
- We need to promote familiarity with using the ICS safely through actual application in robust exercises of sufficient duration and complexity to give the learner hands-on experience. These would be based on the NIMS event type qualifications (e.g., Type 5/4 event IMT, Type 3 event IMT, Type 2 IMT, and finally Type 1 IMT). This competency validation process would be repeated at least every 5 years unless the student can provide documentation of serving on an IMT of the event type he/she wished to retain qualification in.

The idea of identifying key attributes important in an individual's work performance goes back to Roman times when this approach was used to select good soldiers. Even after acceptance into service, new recruits were placed with more experienced soldiers to be mentored and learn their skills before actual battle. Emergency Managers fresh from school or incident commanders who've recently passed their ICS 300–400 courses are not ready for battle, yet we constantly expect them to function.

The modern interest in competencies developed during the 1970s when McClelland (1973) suggested that organizations may be better served by focusing on individuals' competence at tasks rather than their intelligence or scholastic aptitude. Considerable literature exists regarding competence. NIMS/ICS compliance appears to be more about a numbers game than producing competency. Fair enough that assessing competency may be too much of a mandate for the federal government.

The organization is the one responsible for determining who meets the qualifications to be an incident commander. When incidents are large enough to require Unified Command structures or event Unified Area Command organization with local, state, or regional EOC support by an Emergency Manager, who is ensuring competency? Its 'one thing to have individual achieve task competencies, it's another hurdle to have good team competencies', especially when a major event may be the first time this particular group of individuals come together.

Hoffmann's (1999) review of the literature highlights that competencies have been defined as assessing either:

- Observable performance;
- Standards for performance;
- The underlying attributes of a person that lead to performance.

The authors have taught hundreds of ICS 300 and ICS 400 classes. Team performance improvements over a week's time are observable. The challenge is that these same groups may not be the ones who are in positions of managing the next major event. Major events usually mean higher visibility within and organization of often supervisory rank becomes the team member selection factor, not demonstrated competency in incident management.

Both team and taskwork competencies consist of a range of knowledge, skills, attitudes, and other characteristics. However, the key difference is that taskwork competencies support the achievement of individual-level tasks while teamwork competencies support interdependent tasks and thus the functioning of the team. Therefore a successful team is most likely to contain personnel with the requisite individual expertise required for the tasks at hand and the skills and appropriate attitudes to contribute to the interpersonal and social aspects of teamwork. The Emergency Manager may be functioning as a member of a team or in the unenviable position of trying to orchestrate or support a team performance.

In terms of the competencies required for incident management, there have been some interesting studies. Crichton and colleagues compared previous research findings from emergency services and military settings with interview data collected from an IMT called in to manage a major industrial accident on a Gulf of Mexico oil rig (Crichton, Lauche, & Flin, 2005).

The authors identified 5 competencies:

1. Situation awareness
2. Decision making
3. Teamwork
4. Leadership
5. Communication

Australia uses an Incident Management System similar to the United States. The Australasian Inter-agency Incident Management System (AIIMS) provides a common operating framework that IMTs use to manage these emergencies. The Australians use the term Incident Control for Incident Command, but the job responsibilities are identical. A 2006 research project identified the competencies for Incident Management Team (IMT) personnel managing the most complex types of bushfire incidents.

Experienced IMT members were surveyed on the skill sets they believed were necessary for successful IMT performance at the most complex events.

- Key Competencies for IMT Personnel
- Interpersonal and communications skills
- Disciplined

- Knowledge and application of Incident Management System (IMS)
- Management skills
- Leadership
- Decision-making ability
- Flexible and adaptable
- Analytical thinking and problem solving
- Calm and level headed
- Situation awareness
- Technical expertise

Research participants described knowledge of the IMS was an important factor in events that were managed well. When describing incidents that had been managed poorly, lack of IMS knowledge was ranked higher as a reason for the poor performance. Several interviewees commented that it usually takes somewhere between 4 hours and up to two shifts to get "back into the swing" of incident management. This observation suggests that training exercises may need to be longer and more intensive than those currently offered by many agencies.

Our exercises lack effort

Originally the Homeland Security Exercise and Evaluation Program (HSEEP) surfaced in 2002, with revisions in 2007 and 2013. The concept was intended to be a capabilities and performance-based exercise program that provided a standardized policy, methodology, and terminology for the design, development, conduct, and evaluation of all exercises.

Prior to the HSEEP Emergency Managers were conducting exercises using a similar process of determining objectives and structuring play to allow the activity to occur and developing an AAR. One of the authors participated in the Quakex97 earthquake exercise in 1997 the Pacific NW (Fig. 3).

This was a joint local, state, and federal medical exercise based on a Cascadia earthquake event. The same author participated as an evaluator for Cascadia Rising 2016. In reviewing the AAR produced in 1997 with the 2016 exercise AAR, it became apparent that in close to 20 years very little had changed. The next Cascadia Rising exercise is in 2022. Will there be much address in the previous existing plans of improvement?

As a human, it's accepted that exercise once a week will maintain strength and twice a week will let you improve. Are we really "exercising" our programs and capabilities or are we just going through the motions once a year, or in the case of a major event like Cascadia, once every 5 years? One of the authors spent significant time in Lima, Peru learning about their disaster preparedness efforts since they face a similar

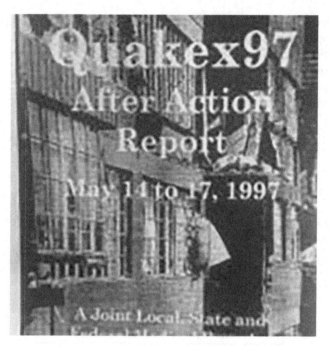

FIG. 3   After Action Reports and Plans of Improvement need to be acted on the actually see improvement.

earthquake/tsunami event. Lima holds national exercises four times a year. Exercises are held at all hours. I'll let the reader draw their own conclusions about which nation appears to take the threat more serious and which nation can expect better performance from both the public and response agencies.

Perhaps we should reconsider our terminology when we refer to our process of training and testing of plans and procedures as exercise. Exercise by definition is an activity requiring physical effort, carried out especially to sustain or improve health and fitness. A good number of emergency preparedness exercises I have observed over the past 30 years don't include the necessary components of resistance, weight, and the exertion of energy. How much mental and physical muscle memory can we really be building? On one LinkedIn post by an author they mentioned that exercises should make the participants work. A new Emergency Manager responded that the author didn't sound like a fun person and it's the wrong approach. Maybe it is, but I can't imagine going to see a personal trainer and asking them for a once a year fun workout that would yield results.

Working out without the use of appropriate weight, repetitions, or resistance is wasting time. When the time comes to need to use your muscles or endurance, you'll be embarrassed. Same goes for our emergency preparedness exercises. If we pretend to work out, we'll be embarrassed by our lack of performance when the real thing happens. We have AAR after AAR that indicates we are not improving as a nation over time.

Is part of the problem of one-off emergency preparedness exercises which includes pretending as opposed to actual performing a skill or decision-making process be accomplishing anything? We take the time to document the exercise and fill in the Exercise Evaluation Guides (EEGs) to develop the AAR and plan of improvement. Then we put away the AAR findings and rest until next year when we exercise again. Exercises should let us work those perishable skill sets, but once a year, with limited personnel participating means we are gaining very little. By the time we exercise again, some people have changed roles, are off shift, or not participating for one reason or another. This means some people may have spans of years before they again practice working on their perishable skill sets (Fig. 4).

Exercising at Cascadia every 5 years seems to almost be not be worth effort. We might pretend we are prepared, but there is no way we build up muscle and mental memory with that kind of pattern.

Suggestions to enhance our exercise effort focused on building and practicing key skills sets could include:

- Exercises for emergency preparedness don't all have to be elaborate and take months of planning. The only ones burning calories for months is the planning team. If you've ever been on a planning team you know exactly what should happen when things kick-off and yet you're constantly amazed at what people have forgotten or fail to

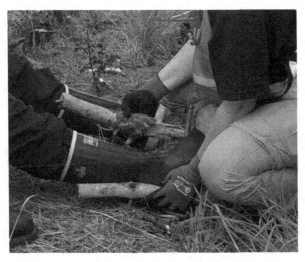

FIG. 4 Catastrophic events mean we'll run out of conventional supplies. How often do we train to improvise?

implement. Your mental fitness is great but the people who have not been exercising their brains are too overweight mentally for a healthy response. One way to get more people to burn some calories regularly is to have mini no-notice "what-if" exercises around the lunch table, coffee machine, or even while driving. The more you exercise the more you can cut through the excess mental fat surrounding what actions you expect people to take.

- Look back over the last 10 AARs from your organization's exercises and I'm sure you will see evidence of mistakes repeated. Focus on these areas over the next year with a smaller and more frequent exercises. Reassess if some of the problems are mitigated or prevented all together by the end of the year. At times the federal grant deliverable process contributes to us failing to ever get good at anything. Changing federal grant guidance wants organizations to focus on exercising X,Y,Z when A,B,C continues to be a problem. It's difficult to build real capacity and skill sets when we're constantly changing our exercise focus.

- Over the author's careers they've come to realize that one of the worst thing organizations can do is not adequately prepare their staff for predictable events and on the specific actions you want them to take. Exercise programs are not for entertainment or creating documents. Participants should feel they have learned something new or reinforced existing knowledge, skills, and abilities. The last thing you want personnel to experience is a sense of doubt and second guessing their actions after a real event because they didn't get to practice. Emergency Managers must hold the lines and facilitate exercises that allow participants to regularly demonstrate competency in critical skill sets.

- The authors have seen too many organizations exercising when they have adequate staff, daylight, equipment, and ideal conditions. Sometimes they will pretend they have adverse conditions, but it is just not the same as trying to work when cold, wet, hungry, and tired. The world changes as night falls and organizations that focus on daytime exercising will find themselves struggling in no time at all in a real major event. Adding the unpleasantness of darkness, wind and rain, and staff shortages during exercises helps build in realism for that one day when someone's life will be dependent upon the decisions and actions of others. Exercises don't have to be elaborate to build endurance. Often just focusing on the simple things under adverse conditions is enough.

- Nothing builds team camaraderie like overcoming an obstacle collectively. One of the authors is known for creating challenging situations as part of exercise evolutions. For example, in one

EMS-Special Operations course the students had a long day of classroom and hands-on skill sessions. They were told to report to a muster point after their last skill session. Most students assumed the day was over and it was just a quick hot-wash. Instead of being released for dinner they were given an exercise scenario with a simple mission. They were to stealth into the military operations in urban terrain (MOUT) site and find their box dinners which were scattered throughout the village. The only hazard would be not being seen by a small hostile force that was conducting random patrols. Rules were easy, no one ate until everyone ate and if anyone was compromised, they had to start over from the beginning. Participants needed to develop their IAP, demonstrate team work and patience even in the face of being tired and hungry, not to mention the weather was windy and wet and waiting until dark to deploy was the only safe option. Participants included civilian medics and one ED physician supporting SWAT units and military medical personnel. It took the team until around 2 am to have their dinner (Fig. 5).

The physician later commented it was the best meal he had eaten in some time. Another student, one of the military medics attached to a National Guard air unit sought out the author several years later and thanked them. He felt the scenario exercises under adverse conditions and at night contributed to them surviving their tours overseas as a medic.

FIG. 5   Happy graduates of civilian EMS Special Operations course.

Exercise like you mean it, sweat and curse and you can talk about it was later how fun and rewarding it was.

Richard Marcinko, a retired U.S. Navy SEAL wrote a couple of quotes regarding the necessity for our training and exercises to reflect real life and prepare us properly. *"The more you sweat in training, the less you bleed in combat." "Pain was their body's way of telling them that they'd pushed themselves to their limits—which was exactly where they were supposed to be."*

It's obvious if we're going to mandate the use of NIMS/ICS to correct deficiencies found in the management of the September 11, 2001 attacks, it's important we are competent in implementing the tool properly. ICS skills sets are perishable and without regular practice with complex events under realistic operating conditions it's clear lack of that knowledge and experience will play a major role in poor incident scene management. Emergency Managers must be competent not just in their ICS skills but also the nontechnical competencies such as teamwork, decision making, leadership, and interpersonal skills to ensure others are competent as well.

Silos will get someone killed

In previous chapters we addressed various decision-making potentials and challenges affecting desired outcomes. In this chapter we view the effects of the various silos, as mental or institutional containers that restrict information gathering, sharing, and the development of additional models of thinking and decision making. These barriers tend to be institutional, mental and even social, and can include our organizations/institutions, disciplines and professions, mental biases, ethnic or religious affiliations, those mirroring our political ideologies, other "tribes," or even those who share some of our and emotional and decision-making biases (Fig. 1).

The primary issue is the need for different perspectives to focus on the one goal of the life-saving and life-protecting mission first, rather than, for example, agency-A following A-playbook, agency-B following-B playbook, and so on. Many of the most important themes that have appeared in other sections of this book will be repeated later, but from the perspective of the damaging and limiting effects of silos.

FIG. 1   For best performance in real event, we need to train with all the agencies expected to be involved.

Organizational/institutional silos

Organizational/Institutional, usually agency-type, silos are probably the easiest to spot. But because they are so evident, and so plentiful, they can easily be overlooked as significant impediments to the useful information gathering, sharing, and diverse explanatory model development of the kind that would have interested Col. Boyd in his quest to keep developing new and varied explanatory models for "nearly everything." Institutional preparedness and response plans, for example, are made by skilled individuals from the various federal, state, and local emergency support functions, and are derived from the traditionally available data sets addressing the regulations in mandated areas. For example, environmental plans based on specific statutes and regulations address air, water, and hazardous cleanups, among other issues; public health and medical systems address much the same issues, and include hospitals, medical staff, and other issues. But the ESF#10 and ESF#8 preparedness and response plans look quite different and are generally made by in-house staff and consultant experts in each area, with not much cross-over expertise and data. After all, if each of the 16 ESF subject areas carried extensive cross-over material, the preparedness, planning, and playbook documents would be as big as old-fashioned phone books and mostly useless for training, exercising, and response. But despite the difficulty of coordinating preparedness and response activities, the results of a lack of routine and ongoing coordination from systemic barricades can be life threatening (Fig. 2).

FIG. 2 Catastrophic events lend themselves to playbooks for coordination of all the players. This gives agencies a chance to ensure priority actions are in sync.

A realistic and practical way to pierce or avoid the stovepipe and to improve decision making and the resultant outcomes is for Emergency Managers in the various agencies to locate the individual(s) in the respective ESFs (agency, nonprofit institution, or even private business) that are needed to assure that nothing "falls between the cracks" or stays barricaded in the silos. Many questions have to be asked, preferably before the disaster events (Isn't this what exercises and training are for?) and they have to be asked in a respectful manner that does not seem to judge or demean the information source. It is here that the importance of existing good relationships with fellow Emergency Managers at state, local, or federal levels; private industry; college/university; and other SME shows its value. Especially if any of these potential information sources has special knowledge or experience in the types of disasters that are being addressed, by preparedness or in the actual response and recovery. This is also the time when a brief scan of *New York Times*, or *Emergency Management* articles available if the disaster that is being dealt with is a large one, or local news newspapers, local blogs, twitter, and so on, and local library historical files if it is a smaller one. It has been said that effective Emergency Managers know lots of things and lots of people, but more important have access to SMEs who have specialties in areas in which they are "generalists." This has not changed as time passes and never will.

Professions and disciplines as silos

There are many ways to wall off and minimize the productive interactions between the various professions and disciplines (the law, accounting/finance, public health and medicine, social work, mental health, civil engineering, management, etc.) and the Emergency Manager that are part of the many groups and individuals that are involved with traditional preparedness, response, recovery, and mitigation missions. Chief among these silos are those based on the way we are trained to think and act in the various professions and disciplines. These learned ways to look at the world are critical, and shape not just how we think about things, but how we react to situations and ultimately, how we make decisions directly resulting in the kinds of outcomes that occur. We should never minimize or just assume that our thinking and decision making will almost automatically result in outcomes that will cut human morbidity, mortality, property loss, and all the other serious issues resulting from disasters. Anything else is just a tool, an interim phase or preliminary to these major missions.

Most professions are consistent in their lobbying (usually successfully) in developing criteria for entry into the profession or discipline, whether we are talking about lawyers, accountants, plumbers, or stage actors. And these entry criteria are "enforced" and protected by professional associations that are charged with protecting the individual members by protecting the criteria that limit entry into the procession or discipline and also enforce by their very existent silos of mind and attitude for the those in the professionals. In classic articles, public statements and writing over 50 years ago, sociologist Nathan Glazer made the then radical statement that professions exist mainly to support the needs of their members, and any public statements they make, for example, in dentistry, or public health, must first be viewed as necessarily self-serving, and then, usually secondarily serving public interests. He later clarified and expanded these views, but for the most part this aspect of them stayed the same.[207] Do we believe this has changed? Here are a couple of examples of how our individual professions or disciplines can drastically affect the major mission, the major outcomes in either direction by providing barriers if not silos to the way the professionals think and react.

Let's start with lawyers and how their legal training and legal practices can adversely affect a major life saving mission if both the lawyers and those who deal with them are not especially cautious and self-aware, never easy things to be. In the discussion that follows, the Office of General Counsel lawyers could not get out of their professional silos, if fact often they acted as if none were there. A DHHS agency (little is served by being specific) had for years used its retirees as well as knowledgeable volunteers to work at large disaster scenes once or twice a year (sometimes more), for one- to two-week missions. The retirees and the volunteers like the extra money but really enjoyed working with their old colleagues and just helping out. The terms of employment were uncomplicated, and payments were made to the former employees as intermittent, at their previous grade and pay. The Office of General Counsel did not pay much attention to the process, in fact assisted in making the process legally "clean." When the former DHHS agency was absorbed into FEMA, things changed. Eventually complex forms were required, and the process became more "formalized." When the agency was again taken from FEMA and was returned to DHHS, there was no provision for hiring former disaster staff under their previous job descriptions and grades, instead they and all other intermittent had to apply for intermittent jobs using the formal, federal job applications website. They were often hired at different grades (usually lower) than they previously had and were now subject to relatively frequent different job descriptions and grading system, and all for employment that was often for just a few weeks a year. At one point during the DHHS-FEMA-DHHS transition, the application forms were nearly 29 pages, though they are much shorter now. Most of the jobs had to do with life safety missions. For years the agency has had too many quality intermittent employees in a revolving door. The problem is usually addressed by another restructuring that changes pay grades, application requirements, and credentials, never quite attaining the stability that poor legal advice has set in place. In fairness, a 2018 change in federal statutes makes it easier to hire former federal employees or retired annuitants for intermittent mission. All of the other complexities for voluntary, intermittent employees are in place, and of course subject to change.

The lawyers that were involved all seemed to have a concept of the law as something that exists clearly in a given situation, as something that has to be followed, whereas in reality there is no such concept. Of course, this was what they were trained to do and apparently their training did not include much about missions other than legal ones. The problem was that the outcomes that were being sought were positive life safety outcomes, not legal ones. Ideally the lawyers assisted in the attainment of the high-quality outcomes all sought, but they most often did not break out of their own silos to do that.

Agency practices are constantly reshaped both by changes in the statutes and regulations or by different interpretations of the existing statutes and regulation. Too often the lawyers forget that their purpose was in helping the agency to more effectively conduct its life saving missions. Conformance to someone's interpretation of the statute and regulations were not the reason they were employed, it was to help achieve the major mission. Complicating matters, lawyers often have higher grades than the line employees who are most affected by legal decisions made in a silo.

And, needless to say, politically appointed managers tend to put safety (of the organization, not the mission) first, with positive agency outcomes second.

Another profession that includes huge amounts of local knowledge and economic and political power in its own vast silo is the real estate profession. A billionaire Brazilian real estate developer plans to develop an 82-unit project where just the penthouse will cost $31 million along the ocean shore in Miami. South Florida officials, meanwhile, have committed to spending $500 million for seawalls; and voters have already approved $400 million in bonds partly to combat issues related to the fact that as the sea levels rise. Some estimates are that over 1/3 of what is now Miami will be underwater in coming decades if nothing is done. But these projects bring jobs, tax receipts, and wealthy residents to the mix.[208] No sensible Emergency Manager can do anything but become bewildered by these conflicting narratives.

Emergency Managers are rarely called upon to give comments to new sets of guiding regulations or administrative processes, though they are subject to them in the field or in the office. There is not much an Emergency Manager can do to counteract poor legal decision made in a silo, but to have developed good personal and professionals' relationships with some of the lawyers in OGC and made comments, as appropriate, when appropriate. Of course, this implies that the Emergency Managers are going "out of his or her lane," and sometimes that trip is required, but it must be done carefully. Further complicating this potential legal silo is the fact that Kahneman, Rosenfield, Gahdhi, and Blaser (October 2016) and a number of others have often found professional judgment is in error, with research showing that man professionals are not consistent in their professional findings, sometimes seem to apply different criteria in the same types of cases, yet consistently have confidence in their own and in their professional colleagues' quality and consistency of professional decisions.

Of course, lawyers and real estate developers are not the only professionals that can get trapped in their own silos potentially damaging the major missions in disasters, protecting lives and property. Another example of how working in a professional silo can damage the major life saving is the extremely complex, multi-solo of attempting to mitigate ocean front property damages from storm flooding, storm surge, and storm winds. Most will agree that mitigation efforts should include not rebuilding homes in locations that are likely to be affected by the next large storm, the selective use of ocean side wetland buffers, select use of sea walls, and generally moving things away from where they are likely to get damaged in the next big storm. The Emergency Manager in these circumstances knows what should be done, but has to face the real possibilities of realtors who not only make their living from commissions on the sale of oceanfront and other properties but tend to look at parts of the world as "saleable" things, homeowners who have most of their disposable savings in an ocean front property, and by the way love the beautiful and valuable view, FEMA insurance and reimbursement authorities, neighborhood associations, village or city governments in various agencies, sometimes state governments in various agencies, local banking interests, and of course local political interests will tend to look at potential mitigation strategies through the lenses of their disciplines and see something as important, and others not warranting serious consideration. A recent publication from the Columbia Law School addresses many of the complexities in ocean front mitigation "Breaking the Cycle of Flood-Rebuild-Repeat; Local and State Options to Improve Substantial Damage and Improvement Standards in the National Flood Insurance Program" the Sabin Center for Climate Change Law, by Dena Adler and Joe Scata, January, 2019, and can be found at http://columbiaclimatelaw.com/files/2019/01/Adler-Scata-2019-01-Breaking-the-Cycle-Final.pdf.[209]

A final example is the agency or institutional fiscal officer who pays close attention to expenditures and who, if not very self-aware, can confuse saving even relatively small amounts of money with the value of the mission being purchased. One of the authors has seen the "value" of purchasing small, $8-dollar texts on the basis of urban planning project review, to newly appointed Plan Commission members who were reviewing hundreds of millions of dollars in a massive planned unit development. These same fiscal "hawks" questioned, but later relented regarding the placement of a subarea, Fire EMS unit adjacent to another massive development. As an aside, many agency or institutional staff who are fired or seriously disciplined are most often damaged for small amounts on their personal travel reimbursement payment vouchers, than for serious errors of professional judgment in substantive areas. Of course, Emergency Managers as well as any wise bureaucrat of staff member will absolutely refuse to file a travel voucher that is not totally accurate and verifiable. If there is an error one needs to complain or just forget the issue, which is usually the wisest course in small amounts. Redoing vouchers, even those in slight error, can take much more time than the value involved if it is small and would be appreciated by the finance professionals reviewing the vouchers. Being kind is often the wisest course.

Philip E. Tetlock may be the ultimate authority on both the values of professional or expert judgment as well its drawbacks. He has spent a career both studying masses of data regarding professional judgment and has also incorporated work from Kahneman,[20] Baron,[210] Klein,[211] Kahneman,[250] Tversky and Kahneman,[212] and a multitude of other experts in various aspects of professional decision making. Quoting Tetlock:

Experts serve vital functions. Their knowledge enables them to perform highly skilled operations—from computer programming to brain surgery to aircraft design—that the rest of us would o well to avoid. Only experts can even begin to craft coherent legislation on complex topics like tax law, health care or arms control. Experts are also, by and large, the generators of new knowledge, in fields from archeology to quantum physics...

But when it comes to judgements under uncertainty—a key component of real-world decision making—expertise confers a less-clear advantage, even in fields like medicine, where one might suppose it essential....

To be sure, there are exceptions in which experts acquit themselves well. Meteorologists and professional bridge and poker players make impressively well-calibrated probability judgements. Veteran firefighters and neonatal nurses can size up situation faster than their rookie colleagues. The pivotal distinction is between experts who work in learning-friendly environments in which they make short term predictions about easy to quantify outcomes and get the prompt, clear feedback essential for improvement—and experts toiling in less friendly environments in which they make longer-range forecasts about harder-to-quantify outcomes and get slower, more vague feedback.[213]

The silos created by our own values, biases, and imprecise mental processes

We have addressed in other Chapters the various biases, decision-making noise, and even emotional threats to accurate decision making. We will relook at some of those same mental limitations on effective decision making, but from the vantage point of their freezing out the normal search for and appreciation for accurate facts, perspectives, and questioning that are usually required for effective decision making, especially in complex disaster-related circumstances. Sociologist Jonathan Haidt[214] summed it up concisely as he observed that when facts conflict with our values, almost everybody finds a way to stick with their values and reject the evidence. In short, we won't ask questions, do our "due diligence" or even question our own irrational suppositions "around the edges" because we are usually unaware of exactly what we are doing and why. Of course, our adversaries or competitors, or anyone else in the room, as we are by now well aware, has a good chance of seeing through our false perspectives even if we can't. Recently, David Brooks, a *New York Times* editorial writer and teacher restated, without using examples from the massive research, the complex relationship between our emotions, our values and cold facts using layman's terms:

(An) unplanned moment illustrated for me the connection between emotional relationships and learning. We used to have this top-down notion that reason was on teeter-totter with emotions. If you wanted to be rational and think well, you had to suppress those primitive gremlins, the emotions. Teaching consisted of dispassionately downloading knowledge into students' brains....Then work by cognitive scientists like Antonia Damasio showed us that emotion is not the opposite of reason.

Emotions assign value to things. If you don't know what you want, you can't make good decisions...Furthermore, emotions tell you what to pay attention tom, care about and remember. It's hard to work through difficulty if your emotions aren't engaged. Information is plentiful, but motivations is scarce.[215]

Stereotypes and generalizations

There are other mental restraints on our abilities to ask questions and otherwise acquire the information and data necessary for good decision making, often hiding right in front of us, where we least expect them. How often have we heard a statement like, "You can't generalize like that." What most people usually mean when they object to "generalizing," is not what that we shouldn't generalize, but that we shouldn't stereotype, which a different issue is. But generalizing and stereotyping are different terms with the problem that in casual discourse people often don't understand or heed the difference, and this makes conversations, especially between experts, or at least more knowledgeable people in an area, and laypeople arduous and exhausting. Stereotyping is an ugly social habit, but generalizing is at the root of every form of science with generalizations as observable statements based in facts and are measure abler and verifiable (Ref. 218, pp. 61–64). So, if our belief is based on a stereotype or worse, a conspiracy theory, no amount of educated discourse or facts can change the mind of those affected. And any of us can be affected under the right circumstances especially if our emotional dispositions or choice are in play. If we are very negatively biased against the federal government, as so many appear to be in the United States can we really accept its recommendations and analysis during floods. We can hope that Emergency Managers can avoid negative stereotyping of any kind and react to situations just as they "see them." But few decisions are unassailable or inspire perfect confidence, especially our own.

The Confirmation Bias as an information gathering silo

In an earlier chapter we discussed the Confirmation Bias, as described by Kahneman[20] (pp. 80, 81) as our tendency to seek data that are most compatible with beliefs we currently hold. This can affect even the Emergency Manager who feels he or she is as coldly rational with disaster decision making as they can be. But the choice of certain state or city emergency management or public health departments as "dependable" or "undependable" may not be based on an accurate assessment of their credibility but instead on how much we have enjoyed working with them or how many of the employees tend to mirror our own values or "tribes." The Emergency Manager should always be aware that his or her choices may not always be rationale, but value or "bias based." These are not pleasant thoughts to entertain and can tend to make our heads swim if we concentrate too much on them, but nonetheless they need to be addressed on occasion. As Kahneman[20] (p. 324) reminds us people (and Emergency Managers) often overestimate the probabilities of unlikely events but also overweigh the effects of those unlikely events in their decisions. If the disaster situation from the past was vivid enough, and serious enough, we will tend to have it appear more strongly in our planning that the risks that reflect it. Of course, the rarity can and will on occasion happen, but, as we stress, only rarely. The point of this discussion is not to reopen a consideration of the Confirmation Bias, which is no doubt the easiest and most frequent way for us to remain in our own value and informational silo, but to resist avoiding the normal need to ask serious questions and gather information from the diverse sources required for effective disaster decision making. And yes, bad decision making during a disaster can get someone hurt, or worse. The Emergency Manager needs to be confident of his and her sources of data, their colleagues, and of the administrative procedures in place, but just a little less confident in their own decision making as being beyond reproach.

Wise thoughts on avoiding silos that are often overlooked

We'll briefly discuss the wisdom of the Dunning-Kruger Effect and the closely related Equality Bias as well as Occam's Razor as important decision-making constraints that can place normal questioning and information gathering and analysis in a fact limiting silo.

In a 1999 study David Dunning and Justin Kruger found that the less competent people were in certain areas, the less likely they were to recognize that. "We propose that those with limited knowledge in a domain suffer from a dual burden: Not only do they reach mistaken conclusions and make regrettable errors, but their incompetence robs them of the ability to recognize it." Emergency Managers, as well as all other "meeting attendees" are reminded that those who know the least are usually the most confident and hardest to convince. And, as we are reminded by Eliot's conclusion "when we do not know, or when we do not know enough, we tend always to substitute emotions for thoughts".[29] Dealing with this common issue is important because the wasted time it generates as well as the potential anger on the part of others can too easily barricade group participants into their own, self-limiting silos. Of course, there are no general solutions, solutions that "fit all" or even most situations because these are the types of problems that first must be recognized before specific solutions can be developed.

Frequently intertwined with the Dunn and Kruger problem of the "untalented" the unskilled or outright incompetent speaking out confidently or otherwise steering a work group into unproductive directions is the what has been called the "equality bias." Speaking of the equality bias, Chris Mooney has written, "…an important successor to the Dunning-Kruger paper has just … come out—and it, too, is pretty depressing (at least for those of us who believe that domain expertise is a thing to be respected and. Indeed treasured)." We will spend a bit more time on the Equality Bias than on some related concepts, because of the obvious effects it can have on Emergency Managers, who are constantly involved in group decision making and in the value of the acceptance of scientific truths (such as meteorologists, seismologists, epidemiologists, etc.).

This time around, psychologists have not uncovered an endless spiral of incompetence and the inability to perceive it. Rather, they've shown that people have an "equality bias" when it comes to competence or expertise, such that even when it's very clear that one person in a group is more skilled, expert, or competent (and the other less), they are nonetheless inclined to seek out a middle ground to determine how correct different viewpoints are.[216]

Driving the point home, sadly, Mooney cited a study that much of the Equality Bias was based on. The study split hundreds of participants into two-person groups so that as the study went on the two members of the small groups got to know one another, their skills and their weaknesses. What happened was that the worst or most least knowledgeable and competent underweighted their partner's opinion (i.e., assigned less weight to their partner's opinion than recommended by the optimum model) whereas the better members of each two-person group overweighted their partner's opinion. Or to put it more bluntly, Mooney reported that individuals tended to act "as if they were as good or bad as their partner" even when they quite obviously weren't. So why did they do that, the researchers, not surprisingly point to the incredible power of human groups, and our dependence upon being good standing members of them. Still, it's pretty obvious that human groups (especially in the United States) err much more in the direction of giving everybody a say than in the direction of deferring too much to experts (Ref. 26, p. 1, 2 in the online *Washington Post* Report, 03.10.2015). The study authors observed that in the United States, at least, that as a society we need to spend more time recognizing the contribution of experts, in such obvious areas as climate change. But overall it shows how human evolution as social groups binds us together powerfully and enforces collective norms but can go haywire when it comes to recognizing and adapting inconvenient truths (Ref. 216, p. 3 in the online *Washington Post* Report, 03.10.2015).

The final "piece of wisdom" highly useful in piercing the silos we can too easily construct, is Occam's Razor (also Ockham's *razor* or Ocham's *razor) from William of Occam, Mid-19th Century*. Occam's Razor means that the simple explanatory theory is the most likely. Occam uses the razor to eliminate unnecessary hypotheses. It can be equated with parsimony in decision making. Without this generally simplifying perspective, the Emergency Manager can too easily view complex disaster situations and retreat into our own emergency silos and develop complex, multifaceted response patterns, when a more direct approach would likely be quickest and most efficient. In Puerto Rico, there were complex data problems, the hill country was inaccessible, the hospital generators were out as well as external power sources, and lots more. Prioritizing this would have taken data acquisition time and data analysis time that was not there. The best approach was to assume the need for life safety issues with transportation, communication, and economic constraints. And the easiest and most important missions were addressed directly, life safety should have been a basic outcome. Of course, since politically selected contractors were given the bulk of the response and mitigation responsibilities, most important things did not get done quickly or done at all.

9

None of us are as smart as all of us

Being smart and being right in emergency management decision making is a task best not done alone. When an Emergency Manager is faced with a catastrophic event, she or he should have a preconceived set of FEMA and other federal, state, and local agency, privately held and nonprofit agencies and entities that can be called upon. Even though FEMA is charged with alerting and coordinating the deployment of many of these assets and related reimbursement issues at least in a major event, the local and state agencies retain the major responsibilities for their areas. An engaged Emergency Manager needs to begin tracking the course of all of these various assets (to the extent realistic), especially in the early disaster response, because almost anything can be forgotten, unwisely deployed, or even lost, at any time, especially during a severe event. For example, as FEMA begins working with state and local authorities and begins providing long-term shelter, food, and water for residents, the National Guard also is available to assist these same jurisdictions by delivering a wide range of services from search and rescue, to clearing roads, drops shipping critical supplies to hospitals on daily runs, and initial clean up and even security (done in conjunction with the Governors). In major disasters other agencies can be accessed quickly, including the US Army Corps of Engineers personnel working with temporary electrical power and generators of varying sizes, inspecting coastal infrastructure and dams. The IRS may be helping taxpayers by extending timelines and by providing other assistance; the Federal Communications Commission will be working with public and private stakeholders to restore damaged networks. In addition, the Federal Motor Carrier Safety Administration may be waiting some regulations regarding commercial vehicles to ease the quick delivery of goods and supplies to affected areas. And the NDMS DMATs can be deployed to augment local primary care physicians and local hospitals and nursing homes that may be having staffing problems or need for assistance from the National Pharmaceutical stockpiles.

The Environmental Agency is available to use local or national contractors to begin addressing the man environmental issues that can accompany catastrophic events. And much more, including the economic stabilizations that can come from using now unemployed local residents for assistance in some of these tasks. In Hurricane Matthew's response and recovery dozens of federal agencies participated. A listing is available on the Pew Charitable Trust cite as follows.[217]

Summary of Federal and state efforts and entities that received additional funding

Federal agencies	North Carolina agencies
Active in affected states	Active in communities
Corporation for National and Community Service	Department of Agriculture and Consumer Services • Animal Welfare Section
Environmental Protection Agency	Department of Commerce • Division of Employment Security
Federal Communications Commission	Department of Environmental Quality • Division of Waste Management • Division of Air Quality • Forest Service
US Department of Agriculture	Department of Health and Human Services • Division of Social Services • Division of Mental Health, Disabilities, and Substance Abuse
US Department of Defense • US Northern Command • US National Guard • US Army Corps of Engineers	Department of Natural and Cultural Resources
US Department of Energy	Department of Public Instruction
US Department of Health and Human Services	Department of Public Safety • Division of Emergency Management • Division of Law Enforcement • Division of Adult Correction and Juvenile Justice
US Department of Homeland Security • Federal Emergency Management Agency • US Coast Guard • US Citizenship and Immigration Services	Department of Transportation • Division of Motor Vehicles
US Department of Housing and Urban Development	North Carolina National Guard
US Department of Interior • US Geological Survey • US Fish and Wildlife Service • National Park Service	Wildlife Resource Commission

Summary of Federal and state efforts and entities that received additional funding—Cont'd

Federal agencies	North Carolina agencies
Active in affected states	Active in communities
US Department of Justice	
US Department of Transportation • Federal Highway Administration • Federal Aviation Administration • Federal Motor Carrier Safety Administration	**Included in Disaster Recovery Act of 2016**
US Department of Treasury • Internal Revenue Service	Department of Agriculture and Consumer Services
US Small Business Administration	Department of Commerce
	Department of Environmental Quality
Included in congressional spending legislation	Department of Insurance • Office of State Fire Marshal
National Aeronautics and Space Administration	Department of Public Safety • Division of Emergency Management
US Army Corps of Engineers	Housing Finance Agency • Housing Trust Fund
US Department of Agriculture	
US Department of Housing and Urban Development	
US Department of Transportation • Federal Highway Administration	

In all, twenty-three federal agencies participated across five states, with more than that many state agencies participating. The many private businesses, nonprofit institutions, and schools and universities that participated must also be approached, have their assets and actions coordinated, and most of all "be kept in mind," since many can self-deploy, as the Red Cross. A small sampling of these kinds of entities follows:

• For profit business and their associations (e.g., downtown retail associations, The Rotary, local Chambers of Commerce, AT&T, Sprint and other communications firms, and many more).
• Local and State water purifications, sewage, and power utilities
• Local school districts, public and privates, universities, and junior colleges
• Private nonprofit institutions such as Red Cross, the Southern Baptist Food Services, Goodwill Industries, the American Legion, Veterans of Foreign Wars, etc.

Of course, the level at which the Emergency Manager is working at will determine his or her levels of involvement. One of the more basic decisions is, after appropriate staff are monitoring and reacting with these individual and groups is determining how will this all be tracked. Choices are, "in electrons" on physical boards and posters, or realistically, on both. Whichever modes are chosen to engage with these assets and track that engagement the Emergency Managers are likely the only individuals (or set of individuals) doing this task, which grows more complex as response gives way to recovery. Sometimes the Emergency Manager is fortunate that it all seems to fall into place, because many of the individuals involved with the disasters are expert in the area in which they are serving, though on occasion that is not the case. Steering a course between being basically uninvolved, or more correctly "overwhelmed," and attempting to be too "controlling" or a hopeless "dabbler," is never an easy set of choices. But the point is well taken that the skilled Emergency Manager needs to respect the individual and group wisdom and talent of most of those also involved in response and recovery (Fig. 1).

Now that we have recognized the value of the various agencies and institutions who will be participating in most disasters, we must recognize a real limiting factor in discussion expertise and talent, of the kind that an Emergency Manager must gather around himself or herself. For a variety of reasons, all of them mostly invalid, too many people are now denying the value of expertise, experience, and talent in an area, even if it has been gathered painstakingly and over long periods. It is true that behavioral economists and social psychologists like Kahneman, Tversky, Thaler, and many others have rightly pointed out the frequent weaknesses in our individual decision making, and even in professional decision making by experts, none have denied the value of knowledge, credentials, and talent.

FIG. 1 A wise emergency manager listens to those around him since none of us are as smart as all of us.

Those who place personal "instincts": or Google research is not using these scholars as their rationale for feeling empowered to reject expertise.[218]

Technology and increasing levels of education expose people to more information than there has ever been before. These gains, however, also fuel a surge in narcissistic and misguided intellectual egalitarianism and derails debates on numerous issues. With only a quick trip through WebMD or Wikipedia, average citizens believe themselves to be as informed as doctors and diplomats. All voices demand to be taken with equal seriousness, and any claim to the contrary is dismissed as elitism. Nichols shows how this rejection of experts developed: the openness of the internet, the emergence of a customer satisfaction mode in higher education, and the transformation of the news, industry into a 24-h entertainment machine. Paradoxically, greater democratic dissemination of information, rather than producing an educated public, has instead created an army of ill-formed, angry citizens who denounce intellectual achievement. Nichols has taken pains to observe that his book was not initially written about a political leader, political party.

Nichols observes that his thoughts have been praised by doctors, lawyers, and teachers as well as by carpenters and plumbers. He also deals with many thorny issues regarding overconfidence in one's decision making, such as the elderly's propensity to value their experience-based opinions over all else, as well as youth's confidence and tendency to challenge things as part of growing up, and of course America's long-standing egalitarian feelings about "eggheads" (Ref. 214, pp. x–xvii). But taken together, he makes a powerful case that expertise is too often challenged by those who don't have enough of it. Will this relatively new phenomenon affect the 'emergency Manager, yes, it certainly seems as if it will, and under the most trying of circumstances during and after disasters.

Global approaches to disaster management

Unfortunately, the international response to disasters is convoluted, at times chaotic, and always complex. Every country has its own hazard profile, vulnerability, fluctuation, and evolution or demise of emergency management systems, as well as unique cultural, economic, and political characteristics. International law is complicated and if details are not fleshed out prior to need, movement of aid across international borders can fail in dramatic fashion, for example, recent Venezuela situation. Even in the United States laws impacting humanitarian aid can stop otherwise helpful medical assistance, as occurred when the offer of doctors from Cuba was refused during Katrina. Each of these influences the country's interaction with international disaster management agencies and organizations. And all the while, disaster management as a practice and a profession is rapidly expanding and improving.[234] The preparedness-response-mitigation-recovery cycle long used in the United States, Europe, Japan, and other advanced nations has now been adopted internationally, with the clarification that disaster relief, a mainstay of addressing disasters in the developing world, is covered mostly under mitigation, with some facets appearing in what would normally be recovery, though strict definitions cannot exist (Fig. 1).

Over 50 years ago Professor of International Law and Policy, H.B. Jacobini[235] stated the obvious, but then almost radical notion that "International Law is International Politics." Not necessarily the rough and tumble politics in the Chicago sense, of the "Don't Make No Waves, Don't Back No Losers" (Milton Rakove, Indiana University Press: Bloomington and London, 1975), but close. In International Politics, quoting Bob Dylan, "Money doesn't talk, it swears." Complicating the already expected, difficult internal political situations of states needing international relief to recover from significant disasters are the political realities that exist among international organizations and the various states providing and needing assistance. And even a further hurdle, if that seems possible, the Emergency Manager serving in the area of international disaster response

FIG. 1   Japan has taken steps to mitigate impact from tsunamis yet still suffered greatly in the 2011 event.

relief must consider the growing power of states like China and how it can make "client states" out of those it provides aid (and sometimes loans) to. Of course, China is but the most recent powerful state that has mixed altruistic relief with political "IOUs."

Coppola (*Introduction to International Disaster Management*, 2015, pp. 701, 2) stresses the importance of differentiating between recovery and development; in many cases it has been the poor condition of some of the societal infrastructure that led to their vulnerability in the first place. It is the reconstruction of the buildings, roads, bridges, factories, utilities, and other components of national wealth that would allow society to function at all; postdisaster recovery efforts will be marked by the difficulties that external organizations have in deciding to which standard that recovery will strive for, as well as how to integrate long-standing developmental goals into these efforts. Coppola (*Ibid.*), p. 700) categorizes early warning systems as expensive, complex, and requiring active maintenance, making them a luxury for the mostly wealthy nations. He observes that when the US government and other industrialized nations were fully aware of the impending tsunami hours before December 2004 and September 2009 but was unable to turn that information into action because of a lack of early warning systems infrastructure in the affected nations (Fig. 2).

Many of the issues that will be addressed in this chapter have been introduced in previous chapters. Some will be expanded or interrelated in recognition that global climate change, growing technological sophistication, increasing urbanization, and growing overpopulation continue to

FIG. 2 Lima is an ocean front city that has taken clear steps to warn citizens about the dangers of tsunamis and measures to mitigate human deaths.

affect the number and intensity of catastrophic disasters overwhelming individual and even clusters of nations:

- Global climate change does not just affect flooding, forest fires, draughts, deadly heat waves, hurricanes, tornadoes, winter storms, and the interrelationships between many of these; it can also affect whole populations and can result in or intensify mass migration, wars, famines, and epidemics, all of which can be utilized by terrorists or despotic governmental actors.
- Growing technological sophistication and social media can be a dual edged sword presenting some enhanced tools for addressing disasters and serious emergencies, as well as growing possibilities for causing them. Unfortunately, the world's growing need for personnel and assets, data capacities, and overall coordination mechanisms will require strong, flexible, well-resourced international systems that for the most part, do not yet exist. But many individual international assets do exist and have improved for decades. The rapid progress toward integrated international systems and the huge need for them require us to continually track progress (or lack of it).
- Populations are concentrating in urban centers throughout the world. Between 2008 and 2010, the world shifted to a majority of urban populations as urbanization rates topped 50%, to reach a projected 70% by 2050. Risk has been concentrated along with populations, with land pressure requiring the poor to settle in undesirable, often dangerous parts of urban centers. These concentrations also focus urban wealth and infrastructure so that disaster situations can have magnified effects on national well-being (Coppola, *Introduction to International Emergency Management*, 3rd ed., 2015, p. 207).
- Overpopulation, especially in already poor nations is most often accompanied by the degradation of the environment; huge food growing and consumption problems; political instability; and massive, destabilizing migrations. On the other hand, as the world grows increasingly educated and more affluent, as increasing globalization persists, these trends may reverse themselves with growing affluence being accompanied by smaller family sizes, if global climate change does not eventually degrade the world's economic capacity to grow. This is just another intersection of fields in which the international Emergency Manager must maintain a good knowledge of the many areas of expertise, at least ostensibly, outside of the boundaries of a more traditional, single-nation based emergency management practice (Fig. 3).

The following discussions will relate to support from big nations to small ones, small nations to small nations, and to the less probable, but still occasionally necessary, support to big nations from many nations.

FIG. 3   Emergency Managers must constantly find common ground among countless stakeholders.

Emergency Managers must place themselves on this continuum and plan accordingly, to the extent that they can. As mentioned earlier in this chapter, the Emergency Manager must take time to understand international aid law and plan accordingly.

The central issue for the Emergency Manager is that a balance must be struck between the rights of the State as a sovereign and the rights of the affected population. Consider the situation where the Emergency Manager's jurisdiction is at risk of a catastrophic event that will overwhelm domestically available resources. International organizations and countries may offer to provide aid to serve affected populations. The question becomes what can the local Emergency Manager do lawfully if such support could potentially meet their needs but also might entail the compromise of national sovereignty? Exactly where the line is drawn is a complex question.

In the 1990s a global consensus emerged that better preparedness, response, and relief cannot fundamentally minimize the risk of natural, technological, or human-induced disasters. It has become understood that clear analysis of the root causes of risk is needed, along with pragmatic approaches geared to anticipating and reducing, to the extent possible, the specific needs of highly vulnerable people to disasters (Ref. 1, p. 259). Many mark the United Nations' declaration of the International Decade for Natural Disaster Reductions (1990s) as the start of the world's recognition of both the need and the possibility of an international response to major catastrophic events that overwhelm nations. During this period the United Nations voted to create the International Strategy for Disaster Reduction (ISDR). Leading researchers have found that:

> ...vulnerabilities precede disasters, contribute to their severity, impede effective disaster response and continue afterwards. Needs, on the other hand arise out of the crises itself, and are relatively short terms. Most disaster relief efforts have concentrated on meeting immediate needs, rather than on addressing and lessening vulnerabilities.

If Emergency Managers were applying these principles they might add that these are the reasons that effective disaster mitigation is so important, but so often underemphasized (Ref. 1, p. 259, citing Mary Anderson and Peter Woodrow[236]). It can also be added that recovery and mitigation (as have been stressed many times) are at least potentially much longer, more complex and costlier than preparedness and response. And even among the most well-intentioned, experienced, and knowledgeable Emergency Managers, vast parts of the future can too easily be "put off" for later consideration, until situations are less complex and time sensitive.

Before we begin to delve deeper into international disaster response and relief methods and organizations, it is instructive to view an extreme example of international disaster relief organizations that have focused on one, very needy country. Right before and during the disastrous earthquake in Haiti in 2010 there were an estimated 10,000 Non-Governmental Organizations (NGOs) working there, yes, 10,000. With its mixture of extreme poverty, disorganization, and proximity to the United States, with its huge array of altruistic relief organizations, it should be no surprise that Haiti has the highest saturation of NGOs of any country in the world. In fact, some Haitians often call their country "The Republic of NGO's."[237] Before dismissing this figure, realize that 10,000 NGOs, with perhaps 7 or 8 employees on average, would only constitute 75,000 or so out of a population of 10 (now over 11) million, before the earthquake. The earthquake killed about a quarter of a million, injuring another 300,000 in a little country already beset with problems of governmental and societal graft, occasional NGO self-dealing, and an overall lack of both effective NGO transparency and coordination (Fig. 4).

In the United States, there are 1.5 million NGOs and 10,000,000 worldwide so, despite Haiti's "leadership" in these areas, it is not shockingly out of step. Of course, NGOs serve a variety of other useful and desirable functions in addition to disaster relief and health care; they serve in the arts, in culture concerns, political advocacy, the environment, minority

FIG. 4 Countries with poor infrastructure not surprisingly do poorly when it comes to disaster preparedness in most cases.

rights, and various kinds of educational and other advocacy issues. But though most have good intentions, various levels of resources, and skilled professional staff and volunteers, this is no guarantee of their participation in effectively coordinated international response and relief operations.

Response in an international disaster context

The myths and realities of human response to disasters discussed here are frequently based on US and Western nation-research and experiences. Considering this, it is reasonable to question if the same kinds of response patterns would be present in another country. Do people in other countries respond in the same, pro-social manners as they do in the United States, with its heavy reliance on informal social networks, including family and friends? Does this similar response pattern exist in international responses? There are basic similarities in international responses, but there are significant differences too.[238] Seeing and judging both the similarities and the differences requires not just cultural and political sensitivities in the Emergency Manager but the wide knowledge and skill sets that are constantly being alluded to here. Stated simply, international disaster response and relief are harder than responses, recoveries, and mitigation done in a single country, in most instances. The 911 WTC attacks, Hurricane Katrina, and the 2017 Hurricane Maria response/recovery/mitigation in Puerto Rico are some of the outliers that prove the "rule."

In the wealthier, most developed countries internationally such as the United States, Canada, Australia, Japan, the European Union nations and others, disaster impacts are typically, relatively effectively absorbed. While the financial impacts can be very high because of their built environments, the loss of life and social devastations is generally low as compared to that experienced in less advanced countries. In these countries, such as Indonesia, Turkey, Thailand, Sri Lanka, Yemen, and many others who experience heightened vulnerability to disasters, their heightened vulnerability to disasters is related to the fact that:

(1) *substantial portions of their populations already live in extreme poverty* with all of the limitations that implies.
(2) *widespread vulnerable infrastructures.*
(3) *potentially huge death tolls* that will result from poor and vulnerable populations.
(4) *the weak and generally ineffective political structures* and traditions that most underdeveloped nations experience.
(5) *the general lack of early warning systems.*
(6) *the technological hazards that underdeveloped nations are increasingly affected by* as wealthier nations frequently shift their least desirable manufacturing to nations with cheaper labor and weak to nonexistent environment regulations.[238]
(7) as previously described, *the underfunded and often inept disaster management assets,* most often located in a military unsuited and poorly resourced for effective disaster management.

One of the more difficult aspects of international disaster response and relief is understanding and effectively addressing cultural disconnects between those providing and those receiving the needed assets and services. Culture is based on language and the shared meanings which define how people react to each other socially. This implies the need to do your homework about the local culture and understanding the most effective ways to operate. But the external Emergency Manager cannot be expected to have realistic understandings of local cultures and social practices because of what has been learned in a language course or two, which is rarely enough.[238] Courses in world politics, history, comparative government, and international relations would appear to be a good source of the additional perspectives that the international Emergency Manager needs. As an aside, many believe that the State Department's examination to become a formal member of the US Foreign Service to be the most difficult and comprehensive examination given through US governmental sources.

International public health and medical response and relief

While United Nations sponsored international efforts were coming to the fore, humanitarian public health, medical organizations and institutions had already begun their own, organized international coordination efforts many decades before. There are a number of significant public health and medical international organizations that deploy to disasters as well as conduct many other humanitarian relief missions worldwide. Among them are the International Federation of Red Cross and Red Crescent Societies, with 191 independent member societies and millions of volunteers. The American Red Cross society dates back to 1863 to its roots in Switzerland, has approximately 20,000 professional staff, 300,000 volunteers, in 364 local units with an overall annual budget of approximately $2.75 Billion. It has also received blood denotations from 2,85,000. Another smaller, but significant NGO that can be depended on to assist at major disasters is Doctors without Borders, with offices in 21 countries, serving in 70 countries with 36,482 members. Doctors without Borders was founded in Paris, France in 1971 and has a permanent staff of 2000. In 2017 there were approximately 750,000 patients admitted to hospitals, performed surgery on 110, 000 patients, and treated over 10,648,000 patients and all with a budget of $390,000,000 in 2017 (Fig. 5).

The United States does deploy Disaster Medical Assistance Teams (DMAT) out of the National Disaster Medical System (NDMS) but is not normally a major provider of international medical assistance, though hundreds of DMAT staff were sent, for example, to address medical issue after Hurricane Maria in 2017 in Puerto Rico which, though "off shore" is a US territory, though it is often referred to as a "possession." In many foreign deployments, health and medical contractors have been deployed by the State Department instead of NDMS assets, but as noted, the United States is rarely a major provider of medical relief.

FIG. 5   State and federal disaster medical teams training for a Cascadia.

For decades disasters were perceived as unavoidable and only "attributable" to naturally occurring events; in the last 45 years professionals in the health field have begun studying ways to limit the effects of disasters, for example, by the founding of the Club of Mainz in 1976, which would become the World Association for Disaster and Emergency Medicine. The Association hosts the only worldwide hazard database. The disciplines of public health and emergency medicine have both made significant contributions. Subsequently, a growing number of professionals have systematically investigated disasters from a multidisciplinary and multihazard perspective.[156] More recently the approval of the International Health Regulations in 2005 by the World Health Assembly empowered public health officials, infectious disease specialists, and epidemiologists to more effectively manage evolving epidemics with the potential to reach catastrophic levels (*Ibid.*, 59). Of course in these and similar epidemiologic issues, the highly skilled, US Centers for Disease Control and Prevention (CDC) is often central to the mission.

As an integrated, international perspective continues to grow in emergency medicine and epidemiology the need for more systematic and data-driven response also grows along with the problems caused by accelerating global climate change and population growth. The main risk for the future is that the importance of national preparedness will continue to be minimized. Although there is still a need for a global response, it must be based on national coordination with increased levels of resource support (*Ibid.*, 69). Unfortunately, addressing huge, potential future difficulties is not always strength of any democratic nation's political system, which can too easily adopt a short term, "by the election" perspective. Fortunately, this is not always the case, as the world eradication of smallpox demonstrates (in many respects led by the US CDC staff).

International disaster and relief agencies and NGOs

The Corporation for National Service (http://www.disastercenter. com/agency.htm) has, with FEMA, compiled a listing of disaster relief agencies available for national or international disaster relief deployments. Created in 1993, the Corporation for National Service is a public-private partnership that engages Americans of all ages in service:

- **The Adventist Community Services (ACS)** receives, processes, and distributes clothing, bedding, and food products. In major disasters, the agency brings in mobile distribution units filled with bedding and packaged clothing that is presorted according to size, age, and gender. ACS also provides emergency food and counseling and participates in the cooperative disaster child care program.

- **The American Radio Relay League, Inc. (ARRL)** is a national volunteer organization of licensed radio amateurs in the United States. ARRL-sponsored Amateur Radio Emergency Services (ARES) provide volunteer radio communications services to Federal, State, county, and local governments, as well as to voluntary agencies. Members volunteer not only their services but also their privately owned radio communications equipment.
- **The American Red Cross** is required by Congressional charter to undertake disaster relief activities to ease the suffering caused by a disaster. Emergency assistance includes fixed/mobile feeding stations, shelter, cleaning supplies, comfort kits, first aid, blood and blood products, food, clothing, emergency transportation, rent, home repairs, household items, and medical supplies.

  Additional assistance for long-term recovery may be provided when other relief assistance and/or personal resources are not adequate to meet disaster-caused needs. The American Red Cross provides referrals to the government and other agencies providing disaster assistance.
- **The Ananda Marga Universal Relief Team (AMURT)** renders immediate medical care, food and clothing distribution, stress management, and community and social services. AMURT also provides long-term development assistance and sustainable economic programs to help disaster-affected people. AMURT depends primarily on full- and part-time volunteer help, and has a large volunteer base to draw on worldwide. AMURT provides and encourages disaster services training in conjunction with other relief agencies like the American Red Cross.
- **Brethren Disaster Ministries** provides volunteers to clean up debris and to repair or rebuild homes for disaster survivors who lack sufficient resources to hire a contractor or other paid labor. Working with long-term recovery committees, the volunteers stay until the work is done. The presence of these volunteer work teams helps to ease the trauma that is felt in the aftermath of a disaster.
- **The Catholic Charities USA Disaster Response (CCUSADR)** is the organization that unites the social services agencies operated by most of the 175 Catholic dioceses in the United States. The Disaster Response section of Catholic Charities USA provides assistance to communities in addressing the crisis and recovery needs of local families. Catholic Charities agencies emphasize ongoing and long-term recovery services for individuals and families, including temporary housing assistance for low-income families, counseling programs for children and the elderly, and special counseling for disaster relief workers.

- **Children's Disaster Services (CDS)** provides childcare in shelters and disaster assistance centers by training and certifying volunteers to respond to traumatized children with a calm, safe, and reassuring presence. CDS provides respite for caregivers as well as individualized consultation and education about their child's unique needs after a disaster. CDS creates a more favorable work environment for the staff and volunteer of their partner agencies. Through consultation or workshops specifically tailored to each situation, CDS works with parents, community agencies, schools, or others to help them understand and meet the special needs of children during or after a disaster.
- **The Christian Disaster Response (CDR)** works in cooperation with the American Red Cross, the Salvation Army, Church World Service Disaster Response, and NOVAD to enable local church members to become effective volunteers for assignment on national disasters. CDR provides disaster assessments, fixed/mobile feeding facilities, and in-kind disaster relief supplies. CDR also coordinates and stockpiles the collection of donated goods through their regional centers throughout the United States.
- **The Christian Reformed World Relief Committee (CRWRC)** has the overall aim of assisting churches in the disaster-affected community to respond to the needs of persons within that community. CRWRC provides advocacy services to assist disaster victims in finding permanent, long-term solutions to their disaster-related problems, as well as housing repair and construction, needs assessment, cleanup, child care, and other recovery services.
- **The Church World Service (CWS) Disaster Response** assists disaster survivors through partner organizations in the United States and worldwide on behalf of its 35 member communions plus affiliated agencies. CWS deploys Emergency Response Specialists who (1) coordinate and conduct training to assure that its partners can carry on effective long-term recovery efforts when disasters strike and (2) work with its partners in developing and implementing projects that address unmet needs of vulnerable populations.
- **Enterprise Works/Volunteers in Technical Assistance** provides telecommunications and management information systems support to the emergency management community.
- **The Episcopal Church Presiding Bishop's Fund for World Relief** responds to domestic disasters principally through its network of nearly 100 US dioceses and over 8200 parishes. It also sends immediate relief grants for such basics as food, water, medical assistance, and financial aid within the first 90 days following a disaster. Ongoing recovery activities are provided through rehabilitation grants, which offer the means to rebuild, replant

ruined crops, and counsel those in trauma. The Episcopal Church works primarily through Church World Service in providing its disaster-related services.

- **Feeding America** is the nation's leading domestic hunger-relief charity. Our mission is to feed America's hungry through a nationwide network of member food banks and engage our country in the fight to end hunger.
- **The Friends Disaster Service (FDS)** provides cleanup and rebuilding assistance to the elderly, disabled, low income, or uninsured survivors of disasters. It also provides an outlet for Christian service to Friends' volunteers, with an emphasis on love and caring. In most cases, FDS is unable to provide building materials and, therefore, looks to other NVOAD member agencies for these materials.
- **The International Association of Jewish Vocational Services (IAJVS)** is an affiliation of 26 US, Canadian, and Israeli Jewish Employment and Vocational and Family Services agencies that provides a broad spectrum of training and employment initiatives needed in disaster. Some of these specific services include vocational evaluation, career counseling, skills training, and job placement. In addition to providing vocational services, IAJVS is also involved in problems of drug and alcohol abuse programs for the homeless, specialized services for welfare recipients, and workshops for disabled individuals.
- **The International Relief Friendship Foundation (IRFF)** has the fundamental goal of assisting agencies involved in responding to the needs of a community after disaster strikes. When a disaster hits, IRFF mobilizes a volunteer group from universities, businesses, youth groups, women's organizations, and religious groups. IRFF also provides direct support and emergency services immediately following a disaster such as blankets, food, clothing, and relief kits.
- **The Lutheran Disaster Response (LDR)** provides for immediate disaster response, in both natural and technological disasters, long-term rebuilding efforts, and support for preparedness planning through synods, districts, and social ministry organizations. The disasters to which LDR responds are those in which needs outstrip available local resources. LDR provides for the coordination of 6000 volunteers annually. In addition, LDR provides crisis counseling, support groups, mental health assistance, and pastoral care through its accredited social service agencies.
- **Mennonite Disaster Services** assists disaster victims by providing volunteer personnel to clean up and remove debris from damaged and destroyed homes and personal property and to repair or rebuild homes. Special emphasis is placed on assisting those less able to help themselves, such as the elderly and handicapped.

- **The National Emergency Response Team (NERT)** meets the basic human needs of shelter, food, and clothing during times of crisis and disaster. NERT provides Emergency Mobile Trailer units (EMTUs), which are self-contained, modest living units for up to 8–10 people, to places where disaster occurs. When EMTUs are not in use, they serve as mobile teaching units used in Emergency Preparedness programs in communities.
- **The National Organization for Victim Assistance** provides social and mental health services for individuals and families who experience major trauma after disaster, including critical incident debriefings.
- **The Nazarene Disaster Response** provides cleanup and rebuilding assistance, especially to the elderly, disabled, widowed, and those least able to help themselves. In addition, a National Crisis Counseling Coordinator works into the recovery phase by assisting with the emotional needs of disaster victims.
- **The Phoenix Society for Burn Survivors** provides social services and emotional support for individuals who experience major burn injuries, as well as their families. Three-hundred area coordinators throughout the United States give their time to support burn survivors and their families on a volunteer basis. All are burn survivors themselves or parents of a burned child.
- **The Points of Light Institute** coordinates spontaneous, unaffiliated volunteers and meets the needs of the local community and other disaster response agencies through its affiliated network of local Volunteer Centers.
- **The Presbyterian Disaster Assistance** works primarily through Church World Service in providing volunteers to serve as disaster consultants and funding for local recovery projects that meet certain guidelines. This agency also provides trained volunteers who participate in the Cooperative Disaster Child Care program. On a local level, many Presbyterians provide volunteer labor and material assistance.
- **The REACT International** provides emergency communication facilities for other agencies through its national network of Citizens Band radio operators and volunteer teams. REACT teams are encouraged to become part of their local disaster preparedness plan. Furthermore, they are encouraged to take first aid training and to become proficient in communications in time of disaster.
- **The Salvation Army** provides emergency assistance including mass and mobile feeding, temporary shelter, counseling, missing person services, medical assistance, and distribution of donated goods including food, clothing, and household items. It also provides referrals to government and private agencies for special services.

- **The Society of St. Vincent De Paul** provides social services to individuals and families, and collects and distributes donated goods. It operates retail stores, homeless shelters, and feeding facilities that are similar to those run by the Salvation Army. The stores' merchandise can be made available to disaster victims. Warehousing facilities are used for storing and sorting donated merchandise during the emergency period.
- **The Southern Baptist Disaster Relief** provides more than 200 mobile feeding units staffed by volunteers who can prepare and distribute thousands of meals a day. Active in providing disaster childcare, the agency has several mobile childcare units. Southern Baptists also assist with cleanup activities, temporary repairs, reconstruction, counseling, and bilingual services.
- **The UJA Federations of North America** organizes direct assistance, such as financial and social services, to Jewish and general communities in the United States following disaster. It also provides rebuilding services to neighborhoods and enters into long-term recovery partnerships with residents.
- **The United Methodist Committee on Relief** provides funding for local units in response and recovery projects based on the needs of each situation. This agency also provides spiritual and emotional care to disaster victims and long-term care of children impacted by disaster.
- **The Volunteers of America** is involved in initial response services aimed at meeting the critical needs of disaster victims, such as making trucks available for transporting victims and supplies to designated shelters. It also collects and distributes donated goods and provides mental health care for survivors of disaster.
- **The World Vision** trains and mobilizes community-based volunteers in major response and recovery activities; provides consultant services to local unaffiliated churches and Christian charities involved in locally designed recovery projects; and collects, manages, and organizes community-based distribution for donated goods.

Nongovernmental organizations in international disaster response and relief

As even a quick review of the organizations before demonstrates, the number of significant nongovernmental organizations (NGOs) focusing on international humanitarian relief has grown exponentially in past decades and have come to play a critical role in response and recovery by filling gaps left unfilled by national governments and multilateral organizations. Some of the NGOs, such as the International Committee of the

Red Cross have established an international presence as wide reaching as that of the UN have developed strong local institutional partnerships and a capacity to respond almost immediately and with great effectiveness. International and grassroots-level organizations alike have achieved such great success in their operations that major bilateral development agencies (e.g., the US Agency for International Development Office of Foreign Disaster Assistance (USAID/OFDA) and international organizations (e.g., the UN) channel assistance through these organizations as an established component of their ongoing and disaster-related relief efforts. There are several classifications (not definitive categories) of humanitarian organizations, with the following broad categories attempting to address them in an organized manner:

- NGOs motivated by religious beliefs or humanitarian goals that are not supported by and do not represent formal government organizations. CARE, the Rotary International, Catholic Charities, the International Red Cross, and OXFAM are examples.
- Private Voluntary Organizations (PVOs), which are nonprofit, tax-exempt, and receive at least part of their funding from private sources. This classification has become more and more folded into the category of NGOs.
- International Organization (IO), an organization with international global presence and influence. Donor Agencies, private, national, or regional organizations whose mission is to provide the financial and material resources to humanitarian relief. Examples are USAID, the European Community, and the World Bank.
- Coordinating Agencies, associations of NGOs that coordinate the activities of hundreds of preregistered organizations to assure responses with maxim impact. Examples include InterAction and the International Council for Voluntary Agencies (Ref. 2, pp. 352–54).

Dimento and Doughman[239] explain that the politics of environmental policy, for example, at the state levels are quite different from those nationally (or even internationally, with respect to complex NGO participation); this is especially true for climate change: "Contrary to the kinds of political brawls so common in debates about climate change policy at national and international venues, ...state-based policy making has been far less visible and contentious, often cutting across traditional partisan and interest group fissures" (citing Ref.[240]). In California, for example, though some legislation involving motor vehicles face strong industry lobbying, but still is supported by environmentalists, Silicon Valley business leaders, and big cities (Barry Rabe, *Statehouse and greenhouse: the emerging politics of American climate change policy*: Washington, DC, Brookings Institute, 2004, 143). American states may be emerging as international leaders at the very time the national government continues to be portrayed as an international laggard on local climate change (Rabe, 2004, p. xiv).

International disaster response to complex humanitarian events (CHEs)

When a nation's emergency management structure is overwhelmed that event becomes known as an *international disaster*, and involvement of the international community of responders is required. Significant changes in hazard profiles due to climate change and human settlement patterns are leading to an increase in the types of capacity-exceeding events that overwhelm nations and increasingly involving multiple nations, thus requiring the help of the international community and stakeholders to include governmental agencies, international organizations, NGOs, and financial institutions. Rich and poor nations are affected by serious disasters, but developing nations because of their physical, economic, and environmental problems experience more capacity-exceeding disasters. In fact, 90% of disaster-related injuries occur in nations with per capita income levels below $760 per year (IPCC, 2012) while 97% of disasters remain weather related. To compound the difficulties experienced by these poor nations is the fact they are most often subject to the types of internal civil conflict that lead to complex humanitarian emergencies (CHEs); these nations usually are more acutely focused on education and infrastructure, or on military demands rather than on reducing short- or long-term hazards with the result that their capacity thresholds are crossed over much earlier than in more economically advanced nations. In fact, it is the military that are used to respond to most disasters in the less developed nations and are generally not skilled in doing so (Ref. 2, p. 331–33).

Haddow et al.[2] (p. 333) list the elements that the US Agency for International Development (USAID) associates with a nations' experiencing a CHE:

- *Civil conflict,* rooted in traditional ethnic, tribal, and religious animosities, is usually accompanied by widespread atrocities.
- *Deteriorated authority of the national governments* such that public services disappear and political control dissolves.
- *Mass movements of population to escape conflict or search for food* can result in refugees and internally displaced people (IDP). A classic example in 2018 and 2019 was the influx of central Latin American immigrants seeking US asylum.

- *Massive dislocations of the economic systems* can cause hyperinflation and the devaluation of the currency, major declines in gross national product, skyrocketing unemployment, and market collapse.
- *A general decline in food security* can often lead to severe malnutrition and occasional widespread starvation.

Haddow et al.[2] (p. 334) have listed five key issues that must be addressed when responding to international disasters. These include:

- *Coordination*, the most important aspect of the response process because of the sheer number of international organizations that can be involved. The need is especially acute in developing nations that already suffer from CHEs. The fact that neither the United Nations, which is the major coordinating agency, nor any other coordinating body has authority over sovereign nations at a disaster scene places emphasis on coordination; poor coordination will almost universally be the cause of failed or minimally successful humanitarian missions after disasters.
- *Sovereignty of the state* is characterized by territory and autonomy and required a request for assistance. This can be extremely difficult in CHEs that have resulted from civil war, as plagued efforts in Somalia with no functioning government with authority over all its territory.
- *Equality in relief distribution* is characterized by certain groups in need being favored over other groups in need. It can happen anywhere with gender and class bias among the more frequent examples of discrimination in relief distribution.
- *Linking relief and development* is an opportunity to develop policy and practice that not only address disaster-related needs but addresses development, as possible.
- *Political implication*, unfortunately addressing local national political considerations as well as those presented by myriad relief agencies, is a task that is usually more complicated than meeting disaster relief needs themselves. This topic has been addressed in many locations in this book.

United Nations disaster management efforts

In the event of any serious disaster, the 293-member UN is quite possibly the best positioned to coordinate disaster relief and to work with the donors and recipient governments involved in rehabilitation and reconstruction. Up to 70% of the UN's budgets is allocated to development, a major aspect of wise disaster recovery and mitigation. Once a major disaster occurs, the UN Emergency Relief Coordination office heads the international response immediately in the form of food, shelter, medical assistance, and logistic support through a committee of several humanitarian bodies. These include: the UN Children's Fund (UNICEF), the UN Development Programme (UNDP), the World Food Program (WFP), the UN High Commissioner for Refugees (UNHCR), and other associated agencies, depending on the specific needs generated by the disaster. After the International Decade for Natural Disaster Reduction of the 1990s (which strove to shift from disaster response-oriented projects to disaster mitigation) as part of its central mission, the UN initiative has since been updated every decade and has concentrated on the following mechanisms:

- Increasing public awareness
- Obtaining commitment from public authorities
- Stimulating interdisciplinary and intersectoral partnerships and expanding risk-reduction networking at all levels
- Enhancing scientific research of the cause of natural disasters and the effects of natural hazards and related technologies and environmental disasters on societies

The UN Office for the Coordination of Humanitarian Affairs (UNOCA) also monitors the onset of natural and technological disasters. This includes dispatching training assessment teams before disasters strike, as well as the conducting of postdisaster evaluations. It also includes attempts to develop common strategies, assess situations and needs, convene coordination forums, mobilize resources, address common problems, and administer coordinating mechanism and tools (Ref. 2, pp. 337, 338, 344, 345). As global populations converge into more concentrated urban areas, their collective risks amplify to an increasing degree. Many of the resultant disasters, particularly in the lesser developed countries will contribute to development obstacles and regional instability unless trends toward increased multilateral cooperation in disaster assistance are recognized more widely than they are at present. National governments, NGOs, nonprofit organizations, international organizations, and the international financial institutions are vital for both preparation and mitigation of hazard risks, and the response and recovery of actual disasters (Ref. 2, pp. 375).

In his massive and comprehensive, *Introduction to International Disaster Management*, 3rd ed. (Elsevier: New York & Oxford, 2015, p. 12) Damon P. Coppola cites the powerful and optimistic United Nations General Assembly goals set in 2014:

- Reducing disaster mortality by half by 2025 (or by a given percentage in a given period of time);
- Reducing disaster loss by a given percentage by 2025; and
- Reducing disaster damage to housing, educational, and health facilities by a given percentage by 2025.

A cynical response to these lofty goals is not warranted as the United Nations does not have statutory or regulatory authority over any sovereign nation, and operates exclusively with its own relatively modest funding authorities and the extensive, but too widely spread resources of the various NGOs, national governments, and international coordinating bodies. It does what it can with the authorities, complexities, and limited resources it has at its disposal. Too often the UN is subject to unfair criticism that does not recognize the extreme difficulties the UN must overcome to be even partially effective in response and relief. Of course, sometimes the harsh criticisms are true.

The development of national and international disaster research

In recent years disaster research has rapidly spread out from the United States, where much of it had its start, now globalizing and extending to the developing world. For example, research been conducted on cyclones in Bangladesh, flooding and landslides in Honduras, earthquakes in India and Pakistan, and the devastating tsunami of 2004. The international perspective is invaluable because, as has been often stated, disasters in developing countries are more catastrophic and difficult to deal with than in developed nations. Although the attacks on the World Trade Center in New York or Hurricane Katrina caused hundreds of billions of dollars in damages, the deaths were primarily, relatively low impact in a worldwide perspective, whereas deaths in developing countries can and have recently caused death numbering in the tens or even hundreds of thousands. The most established research centers in the United States are as compiled by[238]:

- The Disaster Research Center at the University of Delaware (www.udel.edu/DRC).
- Natural Hazards Center at the University of Colorado (www.colorodo.edu/hazards).
- Hazard Reduction and Recovery Center at Texas A & M University (http://archone.tamu.edu.hrdd/).

Numerous research centers outside of the United States have been established to include:

- Flood Research Centre at Middlesex University (http://www.mdx.ac.uk/research/areas/geography/flood-hazards/index.aspx).
- Centre for Risk Community Safety at the Royal Melbourne Institute of Technology in Australia (http://www.rmit.edu.au/browse;ID=-6ccvow449s3t).

* Crisis Research Center at Leiden University in the Netherlands (http://www.socialsciences.leiden.edu/publicadministration/research/crisis-rersearch-center.html).
* Risk and Crisis Research Center at Mid Sweden University (http://www.miun.se/en/Research/Our-Research/Centers-and-Institutes/RCR/).

In addition to these Centers there are now a number of national (US) and international academic, research-oriented journals in emergency management that Emergency Managers should have a decent familiarity with as listed, with some additions, by.[238]

* *Australian Journal of Emergency Management*
* *Disaster Management and Response*
* *Journal of the American Planning Association* (addresses emergency management selectively)
* *Journal of the Association of Politics and Life Sciences* (addresses emergency management selectively)
* *Disasters*
* *Disaster Prevention and Management*
* *Emergency Management Review*
* *Environmental Hazards*
* *International Journal of Emergency Management*
* *International Journal of Emergencies and Disasters*
* *Journal of Contingencies and Crisis Management*
* *Journal of Emergency Management*
* *Journal of Homeland Security and Emergency Management*
* *Natural Hazards Review*
* *Prehospital and Disaster Medicine*

As leading researchers have learned: "Vulnerabilities precede disasters, contribute to their severity, impede effective disaster response and continue afterwards. Needs, on the other hand, arise out of the crisis itself, and are relatively short-term. Most disaster relief efforts have concentrated on meeting immediate needs, rather than on addressing and lessening vulnerabilities.... The distinction (is) between vulnerabilities, which are underlying conditions, and needs, which are created by the particular crisis...." (*ICMA Emergency Management, 2nd Ed.*, 2007, p. 259 citing Mary Anderson and Peter Woodrow, *Rising from the Ashes: Development Strategies in Times of Disaster.* Boulder, Colo.: Westview Press, 1989, 10).

Matthew E. Kahn[241] has observed international and US urban disasters and has long studied how they can improve or degrade as the recovery and mitigation proceeds. He has made a number of pointed observations:

- *Destruction Often Triggers a Boom*: In the classic example of the Chicago Fire which destroyed 2124 acres and 17,450 buildings, killing over 300 while leaving 99,000 homeless. Afterwards, buildings were erected that had an average of three times the value as those that were lost. Wooden buildings were replaced with ones made of stone, brick, or even metal and the construction boom ended up boosting population from 298,000 to 503,000.[242]
- *But a Federal Government "Jumpstart" is not Always a Free Lunch*: Despite investing tens of billion in the region the population of New Orleans has shrunk from 627,000 to 485,000 in 1960 to 311,000 in 2008.[243]
- *Government Activism Can Put More People at Risk*: The federal government responds to public pressure, often because of strong media coverage as well as from economic pressures. A classic example is the rebuilding of neighborhoods and businesses in areas prone to reflooding. Ideally this would not happen, but economic wisdom is not necessarily a national priority in the United States or in any other nation.
- *What Doesn't Kill Us Makes Us Stronger*: Sometimes a disaster and the threat of future disasters will result in stronger building codes, for example, recent earthquake standards for hi-rise buildings in San Francisco (late, but better late than not at all) or higher resistance to hurricanes in structures built in Florida since the 1996 Hurricanes.
- *We Are Not All in the Same Boat*: Those with the highest levels of affluence and resulting political power will usually come out of disasters much better than those who are poor, and live in more-disaster prone areas in substandard structures. In an earlier section of this book we discussed the many poor residents who lost housing in Houston in the 2017 hurricanes, and who still have not received their full share of FEMA reimbursement.
- *People Migrate in Response to Shocks*: Many environmentalists are worried that in the developing world, climate change will create a new class of people called "environmental refugee," who will migrate across national borders as they seek safer areas featuring the basic necessities of life. One of the roots of the Syrian conflict of starting after 2013 resulted from growing drought conditions and the ensuing frictions involving the ownership and the use of arable lands.
- *People Only Respond to What They Perceive to be Direct Threats*: Dimento and Doughman remind us that despite the growth of research and a scientific consensus, unless the dangers from climate change are perceived to be real and immediate action is required, the public is likely to reject the required and costly lifestyle changes.[244]

Inappropriate international relief efforts

Inappropriate international relief efforts

It is common knowledge that one of the problems with international relief (and US relief) efforts is that the wrong kinds of donations are often sent, usually based on what the donors feel is required, and not necessarily on what the Emergency Manager or the many NGOs and other agencies involved, feel is appropriate. And as always, the highest form of relief assistance is money. Complicating this all are the usual local, national, and even international governmental and organizational political considerations. And, there is always the potential utility in having relief and disaster mitigation efforts merge, to the extent that they can, with developmental assistance in less advanced countries.[238] This all is quite a complex undertaking for the rare, skilled Emergency Manager in less developed nations, or even for skilled Emergency Managers in "sending" countries. Philips, Neal and Webb remind us, "Effective development aids those most at risk from failed programs as well as disasters."[238] In 1983, Fred Cuny, international disaster relief expert said, "The most basic issues in disasters are their impact on the poor and the links between poverty and vulnerability to a disaster... we must address the question of how to reduce poverty...if we hope to reduce suffering and to make a true contribution to disaster relief...." Frederick Cuny[245] concurred with Alan Barton[246] observing that disasters will lay bare the social problems of a society as cited by Philipps, Neal, and Webb.[238]

11

Additional case studies

11. Additional case studies

The use of young military recruits in local/state/federal trainings
and exercises 263
Summary 263
Lessons learned 264

Changes in the military must be recognized and dealt with 264
Summary 264
Lessons learned 264

The unfortunate tendency of federal agencies to repeat big,
well-publicized blunders instead of learning from them 265
Summary 265
Lessons learned 265

Emergency Managers need to extend trust to receive it 266
Summary 266
Lessons learned 266

Emergency Managers may need to improvise solutions not found
in a book (even one as thorough as this one) 267
Summary 267
Lessons learned 268

This chapter includes case studies that were not presented in the text of any of the recent chapters. We focus on decision-making successes, weaknesses, and even the failure to make a relevant decision. In some cases, we focus on just one decision, in others a few. In other cases, we do not focus directly on decision making, but on the patterns or background issues that led to the successful or unsuccessful outcomes. Most of the cases come from the authors' cumulative experiences, though some are from other sources.

Racism, religion, and an unexpected death

Summary (some facts have been slightly altered to shield some participants)

In 2005 one of the authors of this book was preparing to leave an Emergency Operations Center after hurricanes had passed through Texas, parts of which were then in recovery. While at the Ops Center, he was informed by a Texas EMA rep about a busload of affluent, elderly, African American parishioners who were on a holiday bus trip to the mountains in northern Texas. During the trip one of the older parishioners (they were all in the 70s and 80s or more) died, apparently of natural causes. After they resumed their trip the elderly parishioners noticed that there were some clearly racist signs on a few of the houses and along the roadside. Though the signs had nothing to do with anyone on the bus, as mass panic ensued, and the bus driver and trip leaders were afraid that someone else might die, this time from fear and stress-related causes. Hearing of the event the Texas State Police attempted to come aboard and offer assistance. But the travelers were fearful and were not allowing the Texas State Police on the bus.

The Texas EMA staffer asked federal ESF#8 to handle the issue and to try to avoid stress-related injuries. The State was already stretched thin on staff and wasn't interested in working such a thorny issue that would take time and possibly continue long after recovery. The parishioners were elderly, but also affluent and educated, so Congressional inquiries and all sorts of bureaucratic difficulties were discussed. Since the State Police were being barred from the bus, they forwarded the issue to the Department of Health, who forwarded it to the Emergency Operations Center, asking for "The federal health and medical guy." Neither the State nor the local health department staff were anxious to deal with the issue and preferred to have "an outsider" especially one well versed in "disasters" to deal with it. It all happened in a few minutes. At least that part of the system worked and worked very fast.

Quickly, a few responders gathered in front of the ESF#8 Lead to hear what was happening and really, happy not to be involved. The ESF#8 Lead asked the small group of local and federal employees, "Where's the closest church to the bus?" Almost immediately he got an answer, and the phone number was popped a few seconds later. He called the closest church, which was Catholic, a relative rarity in this part of the state. A priest answered, and he was asked to have the State Police drive him to the bus, that would be stopped in a safe area. Appropriate food was ordered and arranged to be delivered to the meeting site and NDMS nurses were immediately dispatched to take the vitals, check meds, and do a quick check up of everyone on the bus. An hour or so later there was a meeting of the bus driver, the priest, the NDMS nurses, and the food with the State Police outside for security. The ESF#8 Lead asked to convey a message to the group's organizers, having him announce that the federal government stood ready to assist them in any way, for the rest of the trip. (This seemed like a bit of an overreach, but seemed like a good thing to say and do.) It all worked out.

Lessons learned

The main lesson was to engage with folks who know more about the situation than you do and take their advice if at all possible. A second lesson included the need to try and understand the panic that was being felt by those in need, and to try and address it by having the most appropriate folks possible to work the issue. The third issue is to think hard and act fast. The fourth is that Emergency Managers need to help people if they are at all able to, even if that requires a trip a bit outside of their lanes. The fifth lesson is that even relatively "small," non-life-saving activities can have a strong impact on people's lives during a response, of any kind. For his successful activities in this "response" as well as for the help that the ESF#8 Lead rendered to the Governor of Texas' Office, the Governor awarded the Lead the coveted Lone Star of Texas flag, as a field award. However, in the heat of the response and recovery, the Flag was never secured, much to the sadness of the ESF#8 Lead. Taking precious time away from the response and recovery just did not seem appropriate. And then, the moment was lost.

Good leadership can sometimes support bad decisions (since emotions for some are still high after more than a decade, some of the facts were shrouded)

Summary

In 2006 the ASPR Regional Administrator (RA) routinely requested NDMS DMAT elements in support of a National Special Security Event mission in a Midwest city. The DMAT elements were to be deployed close to the downtown, where the main events were to occur, but slightly outside of the downtown for ease of access, less traffic frictions, and related logistical and cost-related reasons. The site suggested by the city and accepted by the RA was almost four-thousand square feet in a building that backed up to the main fire department building and was right next to an all-day/all-night place that fire and police frequented on breaks or for breakfast or dinner. Right across a very small street, was a more or less secret location that housed over 200 black clad security support staff, gathered from many local and state police departments. The ASPR staff did not mention this deployment, as we were told to hold the information close. Downtown police were instructed to fill out their time and locations sheets on the streets around the DMAT location, with it in view. There were officers coming by and stopping a few times an hour or more for approximately 10 min. ASPR was in the habit of accepting state, city, or federally recommended locations, and accepted this one. We did ask for at least one-armed security individual to be on site with the team at all times, though we did not expect to have this granted, being that virtually all available police had their time accounted for, and there was no money to hire security staff, as this was not Stafford Act-reimbursed event.

The ASPR RA "handed 'off'" on the ground supervision of the ESF#8 aspects of the event to a RA from another region, who was here to assist, freeing the local staff to view the overall event. Both the out-of-town and local ASPR RAs (and his Deputy, who operated as a small, overall management team) were satisfied with the team security and accommodations, access to local hospitals, coordination with the Secret Service, FEMA, state and local emergency management and public health, and related matters. It was neither surprising nor unexpected that there were some state and local frictions, but nothing that was excessive nor unexpected. High tension, talented Emergency Managers and public health responders sometimes have different views on the specifics of a deployment.

The intel briefings did not indicate any suspicious groups or disturbing trends. That was basically good news, but it never was totally satisfying as the RA felt, and still does, that the really good ones don't make their presence known until the event unfolds. As luck would have it, there was a small demonstration that wound its way around the downtown, mainly on its approved path. Whether or not the demonstrators erred or waivered off the path on purpose, the demonstrators weaved to within one block or more from where our DMAT elements were deployed. Pepper spray was used and some of it wafted back to the deployed team members, where they could notice it, but not to any significant degree. The RAs and the on-the-ground ASPR manager were perhaps two blocks away from the pepper spray and never noticed it. This was the day before the event, and more demonstrations were planned, but none was large, at any rate none were expected to be larger than what we just saw.

The Team's Logistics Officer had a great reputation (deserved) for being knowledgeable, hardworking, and totally dedicated to team efficiency, safety, well-stocked pharmaceuticals, other medical supplies, and all of the other items and equipment that eased a deployment. When he noticed the odor of pepper spray, he began telling the team that they were in a "biological hot zone," he was using hyperbole, but the team knew he felt they were in danger in their present location (as described before they were in no danger). To the surprise of the ASPR RAs watching the whole event unfold, the Team Manager, a highly skilled and credentialed Emergency Room Physician and a noted figure in the emergency medical community believed the Logistics Officer. He immediately called and said he wanted someone with a gun to bunk with the team and began saying he thought alternate locations should be considered. The ASPR RAs reconvened an all hands team meeting at the end of the team's 12-h shift (nights were covered with small placeholder team with all of the rest of the team on immediate recall if necessary; the team was quartered in a downtown hotel a few blocks away, with a view of the activities and even of the demonstrators). At the meeting the Logistics Chief restated his concerns even more vehement, and demanded an armed presence, a consideration of alternative space for the team. In an unwise exchange the ASPR RA restated that their current location shared a wall with the Fire Department's Headquarters, had police constantly driving the perimeters and even doing their reports adjacent to the site, and were as "safe as they could be." The RA did not mention the 200 or more black clad security forces just a few feet away, and did mention that while no armed police were available, an intern with an immediate disaster line was stationed with the team. The RA's talk ended by saying he would gladly "have his wife and daughter bunk with you all tonight." The Logistics Chief exploded that none of this made them safer, and he did not want to be part of a deployment like this. The RA replied that he could see Finance and get his plane back home.

The next day there were more demonstrations but no use of pepper spray. The mission ended uneventfully. A few years ago, on a site visit to the team, the Team Manager again mentioned that the RA had left his team in a dangerous position at the NSSE. He was told about the 200 "buried" backup security forces, and the other elements of team security were repeated. He was also reminded that NDMS DMAT elements rarely if ever travel with armed security, and that NDMS did not even have armed security staff, and always had to secure it, which was never easy. The Team Commander seemed not to hear any of that, and he was, and is, one of the best in the system. The RA was doubly certain of all of this because he was deeply involved with assisting the Team Commander get the team initially developed and certified as part of NDMS.

Lessons learned

The RA seriously misjudged the strength of the respect and confidence that the DMAT, its Team Commander and team members had in their Logistics Chief. He truly believed the team was in danger, and conveyed that emotionally laden position to the team, despite the fact that few outside that room believed that. The RA who also had a strong reputation should have told the team "to trust him" that there were buried assets he could not talk about that were protecting the team, by its very presence. He also could have strengthened his position by mentioning that he was involved with helping the Team Commander organize and certify the team. Also, the Logistics Chief should not have been pushed to leave the site. That was not a wise thing for the RA to do. After a few years the RA and the Logistics Officer reconciled. Disaster deployments can be times of high stress and high emotion; these complex background issues must be constantly addressed by Emergency Managers in their decision making or their decisions may fall prey to the strongest emotions in play.

Ego-driven decision making and behind the scenes power (some facts altered or deleted)

Summary

Just over 25 years ago a much beloved Admiral in the US Public Health Service was with his small staff, at an early and successful deployment of the National Disaster Medical System. The deployment was after a serious hurricane. This was all the more impressive because the Office of Emergency Preparedness had no formal budget and was mostly funded from other pools of staff and funding in various US Department of Health and Human Services with much coming from the Offices of the Regional Health Administrators. In fact, it was rumored that the Admiral would advise then President Clinton regarding disaster deployments and other issue, in his private Presidential quarters, and presumably to secure additional funding. Sometimes the Admiral would deploy NMDS without adequate funding, only to request it after the fact from FEMA or other agencies. Although each deployment was realistic and appropriate, virtually all would agree, but conducted outside of the normal policy and funding grid. The difficulty with the very successful hurricane deployment in question was that the Admiral showing pride in his staff, as well as in his own creation, invited approximately 300 Congressional, federal, and other staff to come through their operation headquarters to view the deployment, which was set in what was otherwise a tropical vacation spot. FEMA, among others, were very angry.

Two weeks later the Admiral's whole staff, then fewer than 20, was called to a side room while at a national exercise and sat and watched an "HHS hatchet" read a list of true, but slanted facts, and then asked the Admiral for his retirement in front of his staff. The staff later found out that another member of the US Public Health Service wanted the job, after seeing its potential and was given it. Later, the staff found out that the individual was previously a floor man for a famous Democratic Senator. The agency was soon at approximately 100 FTEs and was now authorized every year. ASPR developed out of this small, public health and medical disaster agency.

Lessons learned

- Strong pride, or ego, or a combination of both can blind an individual to the bureaucratic war clouds that constantly hover above Washington, DC.
- Often, an individual, even one who may lack social skills can persist, and even thrive if his or her mentor is powerful. The "hidden but strong mentor" is a consistent story in the Headquarters Offices of many federal agencies (and no doubt state and local ones as well).

Six brittle patients needing medical evacuation stranded on Samoa at a closed airport (some facts slightly altered or masked)

Summary

After hurricanes struck the American possession in the Pacific, five or six patients, a couple of NDMS DMAT members, and eight close relatives who would not release the brittle patients, (all on I.V.s) unless they traveled with the patients were stranded at an airport in Samoa. It was night, and the airport had just closed, and to make matters even more complicated, the pilots had flown their max hours and had to sleep for the night. The night shift at the EOC that was covering the hurricane found out about it from one of the NDMS team members who was packaging the patients for evacuation and was now stymied from finishing that task. The call came to OEP to the night desk, then over to ESF#8, and finally to the ESF#8 desk servicing the event. The call came in just before the shift was changing at 8 pm. The Day Operations Chief was informed that the nine members would work the issue and see if anything could be done to make the evacuation happen.

The ESF#8 Lead felt it was a life safety issue and had to be worked on an emergency basis regardless of the potentially slim chance of success. The Day Operations Chief was so intent on following the procedures regarding avoiding overtime and leaving at the end of his preapproved 12-h shift, and any potential threat to his authority, that he replied that the issue could be dealt with in the morning, it was too late to do much at an airport anyway. He said in the morning, he could put an even larger crew on the problems with some flight specialists. The Day Operations Chief then told the oncoming Night Operations Chief as well as the 8 or 9 ESF staff around the table to put off dealing with the issue that he had it covered the next day.

I explained the nature of the brittle patients and how some options had to be immediately worked. The FEMA Day Shift Ops Chief still left. The FEMA Night Operations Ops Chief, CC, looked at me, and shook his head meaning, "Don't worry, we'll figure something out." We opened with me giving a brief, but relatively detailed summary of the whole issue. The FEMA Ops Chief laid out five or six submissions which included the JRMP calling up chain, ending in getting the Secretary of Defense to be awakened in order to waive the pilot's flight requirements, and to keep the dispatching as well as the reception airports open. Lots of calls were made by various folks at the table before these critical calls were able to be made. An appropriate hospital was alerted, patient info was transmitted, a bigger fixed wing was obtained as well as appropriate NDMS staff to both package the patients, to travel with them, and work with the reception facility.

After a few calls the Air Guard staffer got ahold of the Secretary of Defense, himself, and got the airport reopened. Others got the flight hour rules waived and all of the other missions. By 3 or 4 am, the Night Operations Chief was laughing about it all, and telling the group he was probably going to get his "behind kicked" by the Day Chief, when it was over. The calls and meetings took most of the night. By breakfast we were successful, and all went home at the shift change. The Day Chief figured out what had happened, was furious and was taking it out on the Night Ops Chief, who was absorbing the bashing, as we all left.

In this instance the Day Ops Chief did have an experienced wise Emergency Manager advise her to stay on at the EOC until the problems were solved, but she found reasons enough not to hear or accept this as good advice.

Lessons learned

When an otherwise skilled but inexperienced manager oversees an operation that features life safety issues the Emergency Manager, after trying to "be reasonable" must address the life safety issue as quickly as possible, later "apologizing" as appropriate. There is never a time when risking lives or not saving them can play second to administrative issues. These situations are difficult, and rare, but when they surface, they must be addressed with the traditional "Ask forgiveness later" choice.

The young and inexperienced weather scientist who "saved the day" (some facts have been slightly altered)

Summary

A FEMA Emergency Manager was assigned to a State of Texas Emergency Operations Center in Austin during the 2005/6 hurricane season. As the ESFs attended the standard meetings, made or received the standard situation reports the group's attention began to focus on a hurricane that was about one- and one-half days out of Galveston, making its way toward Houston and gathering power from the warm waters as it got closer. We all knew of Galveston's tragic history of the 1900 Galveston Hurricane and storm surge which completely destroyed the city killing 6000 to 10,000 in the process, one of the deadliest weather events in American history. When the hurricane was about 3 h out of Galveston it was a Category 4 and was being projected to possibly reach a Cat. 5 as it made landfall. At about 2 am a very young, and small of stature NOAA Meteorologist walked up to the podium there was a noticeable snicker from the Emergency Managers gathered for the briefing. The young meteorologist heard it too. He quickly announced that our concerns for Galveston were misplaced, that he had noticed "a slight anterior wall deterioration" in the approaching hurricane, that, along with a few other factors" convinced him that the storm would never reach Galveston in any form other than a bad rainstorm, maybe a Cat 1. Most did not stop ordering assets and directing predeployments of assets, though it was late for that. The young meteorologist was right, and we all left the EOC for breakfast, and most did not return, leaving the rainstorm for the next shift.

Lessons learned

The lesson was clear. The young meteorologist should have been paid attention to; he earned the right to give an important briefing because of his credentials and should have been respected. A snap judgment based on physical size and youth was wrong in the worst way possible. Fortunately, assets that may have been diverted to Galveston and Houston were not needed elsewhere, or the biased decision making could have had serious, maybe even deadly results.

The benefits of keeping a small footprint when stepping out of your lane

Summary

While in the EOC in Austin a decade or more ago, while responding to hurricanes, the ESF#8 desk received a special request from the Governor's Office, through the State Department of Public Health. Several dermatologists from an eastern state had volunteered to come to Austin and "help" during the disaster, presumably to practice disaster-related medicine. This presented more than a few problems. In a disaster situation, trained paramedics or better yet, RNs, nurse practitioners or physicians' assistants would likely have more useful experience in emergency situations (knowing, for example, that early in a disaster clean up, puncture wounds from screwdrivers are a big producer of ED visits, etc.). Dermatologists are MDs and that is always good but having them render basic disaster-related primary care would not be a first-choice use. Also, having them sleep in possible austere conditions is not something they likely had in mind when they volunteered. Of course, no one wanted them moving around the disaster scene unprotected. And recall that despite getting NDMS salary and travel reimbursement most will lose money, sometimes lots of it, for their deployment. Even so, the Governor's Office wanted to show respect for these volunteers.

After pondering the issue, the ESF#8 Lead decided to dispatch the dermatologists to the Disaster Recovery Centers where their practice would be directed at Medicaid and uninsured disaster victims, of which there were many. The intent was not to have them give uncompensated (free care) to insured patients, which would be adversely affecting the practices of in-state dermatologists, besides potentially causing political problems. The targets of the volunteer dermatologists were the young as well as the elderly. After a few weeks many cases of teen aged severe as well as not so severe complexion issues were addressed as well as seniors who really should have a review every year or two of the status of moles, and so on. It all worked out as there were no Congressional inquiries and no angry letters from any state professional associations.

Lessons learned

The effective Emergency Manager must be flexible. In many disasters we are working in someone else's state, and that means that the Governor's needs, political or not, need to be addressed as best as we can. Without the cooperation of their offices no response or recovery can be done well.

The use of young military recruits in local/state/federal trainings and exercises

Summary

For those Emergency Managers fortunate enough to have a sizeable military base nearby and a base commandant amenable to having recruits participated in local trainings and exercises, there is a special responsibility placed on the Emergency Manager to try and create the best learning environment for the recruits and not to just use them because they are free and available (though those are good recommendations in themselves). To this extent this section has been drafted not only to recommend making the recruit's experience better, but to recognize the huge benefits that using young, uniformed military recruits can bestow on the Emergency Managers, the recruits themselves as well as on potential federal, local, and state users of young, uniformed recruits.

If it appears that the authors are "biased" in favor of using and providing enriching experiences for these young recruits, which is correct. Our past histories are filled with examples of noticing someone visiting a base and misplacing a wallet, phone, or other object, and having a breathless recruit find the object and then network across a huge base to deliver the object before the person leaves the base; or of seeing the attitudes of the recruits. The almost uniformly great attitudes of the young recruits are evident to anyone who sees or interacts with them. They all were drawn to being part of something larger than themselves, getting valuable training and valuable life experiences.

Lessons learned

They have been stated in the body of this brief section. Using young, uniformed recruits saves resources in federal, state, and local training and exercises. It also implies that the Emergency Manager has a responsibility to make the experience as valuable and enriching as possible from the point of view of the young, uniformed recruits.

Changes in the military must be recognized and dealt with

Summary

As we review past deployments to gather what lessons we can, it becomes very clear that the assets of the National Guard and even those of the "Big Army," which for our purposes includes the Navy, the Air Force, Coast Guard, and all of the uniformed services, often play outsized roles in disaster response and recovery whether or not they are reimbursed by FEMA under the Stafford Act.

A few years ago, one of the authors was made aware of the impending retirement of an in-law who was an 05 in the Air Force, hoping to attain the 06 rank before he retired. The in-law, in looking over his six, 05 competitors said he was afraid it was a "crapshoot," as they were all highly credentialed (as is the practice in the Air Force), had many status assignments across the world and unblemished records replete with awards, commendations, and honors of every sort, and all deserved as best as he could determine. The brother-in-law, for example, was a national expert in logistics and training, had a Master's in Business, spent a year on a detail to Saudi Arabia, taught courses at the Air Force Academy, had many awards and accomplishments and was smart, honorable, personable, and even handsome as he recanted while laughing. He was, as he feared, passed over and had to retire.

As the numbers of enlistees has dropped, so have the opportunities to involve the young soldiers in disaster response and recovery. For example, one of the authors has called an existing base for recruits to assist in large exercises (for example as moulaged injured with learned symptoms and responses), to provide unarmed security and other missions. These missions enhanced the recruits' training experience and were invaluable to the local, state, and federal agencies conducting the exercises and trainings.

Lessons learned

For a number of years, the Officer corps in most of the uniformed services has been shrinking, as has the numbers of noncommissioned officers and enlistees. They have been replaced by sophisticated systems of all sorts while human staffing has generally dropped, and wars and engagements have occupied others. The point of this is that all forms of military assets remain highly valued and coveted assets during disaster response and recovery but have to be viewed in the wider perspective to assure that requests for their uses are realistic and made with some knowledge of the competing demands for these sometimes-dwindling assets.

The unfortunate tendency of federal agencies to repeat big, well-publicized blunders instead of learning from them

Summary

Some have referred to this as the federal "Lemming Tendency" while others have simply called it federal "Monkey See-Monkey Do." A notable example occurred when the Government Service Agency (GSA), which should have known better, was exposed for having made some significant blunders in some of its larger, internal agency meetings. The specifics are not important, what was important was that the errors were big, dumb, and exposed publicly. Of course, of all of the agencies we should expect to be wise in the use of official meeting expenditures, we would expect GSA to be the skilled and the most cautious (following the federal "rule" of "Do whatever you do with full knowledge that it could easily be made public, and if an mistake made is big enough and dumb enough it probably will be). In fairness, the GSA blunder could be the result of just one or two Senior Executives, more likely powerful political appointees, who set the stage for the foolishness that followed, with everyone else just following instructions and figuring it all had to have been properly vetted, so it must be ok.

Instead of learning from the GSA folly, many federal agencies simply banned otherwise highly useful national or quarterly meetings. For example, one agency often had its annual or bi-annual meetings in Las Vegas because it could get blocks of high-quality rooms for as little as $79 a night with free or low cost, large-sized session rooms. Another agency stopped its hugely successful series of annual meetings held in Nashville at the "Grand Ole Opry" resort which had reasonably priced rooms for hundreds of staff and volunteers in a controlled and high security atmosphere. Not having these meetings damaged agency training, cohesion, and possibly even recruiting potentials. After a decade there have been no annual meetings.

Lessons learned

As has been discussed before, federal agencies (and state and local ones, too) would usually rather stress avoiding mistakes than seeking highest quality of outcomes, which can come with relatively high investments of personnel time and resources. This is an undesirable path in the persisting bureaucratic austerity that still seems to be affecting the nation but not one that either benefits the mission, the nation, or the Emergency Manager. Stressing the safe over the best is an attitude that Emergency Managers should avoid as this why most of us have joined the profession.

Emergency Managers need to extend trust to receive it

Summary

In 1989 the National Disaster Medical System, a partnership between FEMA, DOD, and the US Public Health Service had never deployed a Disaster Medical Assistance Team, but had recently arranged for liability protection under the Federal TORT Claims Act, had the volunteer members classified as intermittent federal employees when they were deployed and had developed a number of smaller treatment teams, but only one, fully staffed, 35-member DMAT. There was no uniform, so the members wore a collection of t-shirts from various fire departments, EMS services, or DHHS logo shirts they "found" at various training sessions or meetings. The DOD was to provide air and land transportation to and from deployments of DMATs, DHHS was to supply the DMATs, pharmaceuticals and equipment, and FEMA was to pay for it under its authorized budget. There was no authorized budget for anything in NDMS, yet.

As all of this was coming together, Hurricane Hugo devastated the central US Virgin Island, St. Croix. Department of Defense was approached to fly the DMAT over to St. Croix, from Albuquerque, New Mexico, where it was formed, and even waiting on a tarmac expecting almost immediate deployment to where it was needed. Tom Reutershan, then Director of the program, and then Deputy Director Dr. Stephen Posner made the initial calls that organized the first deployment of a DMAT. At first, Tom Reutershan of the NDMS (Home of the DMATs), was refused an airplane by DOD which said there were none available. Reutershan made a quick decision and used his personal credit card to charter a private jet to carry the 35-person DMAT with all of its equipment to St. Croix, where it was badly needed. It was brave move and placed a lot of faith in FEMA to pay for it all, but lives really were at stake, and it was Reutershan's responsibility to deploy needed assets to St. Croix.

Lessons learned

It was fitting that a system designed to deploy potentially life-saving assets was initially deployed based on Tom Reutershan's faith in FEMA's figuring out how to reimburse him for the private jet costs he had put on his personal credit card. St. Croix public health and emergency management staff were told to expect the deployment of an emergency services team to address their critical medical needs. And FEMA had faith in the nascent NDMS to deploy assets to St. Croix as soon as possible meeting its obligations under the joint agency agreement that formed NDMS.

Emergency Managers may need to improvise solutions not found in a book (even one as thorough as this one)

Summary

At the height of a 500-year flood event (federally declared disaster in multiple states) the author was dispatched as the ESF-8 representative for the state health to conduct a first-hand assessment of a rural community that third-hand reports suggested was in need of assistance. There had been no direct request from the community or the county Emergency Manager for this support. One interesting aspect of this field deployment was that there was no plan for this type of activity, it would have to be make it up as you go mission. The author had taught an EMT class in community years ago, had a four-wheel drive vehicle with a decent communication package.

Power had been out for several days and as I got closer to the community, following the road along a small river, I noticed a debris field 15 ft up in the trees. A lot of water had been through here. Finally arriving in the community, the author found that the survivors were too busy trying to sustain lives to try and contact anyone, there was no power, there was no form of communication available. The volunteer fire chief's home had been washed away but they were trying to manage the response. He took the author to a church that had been established as a shelter and kitchen. One of the first disturbing activities was watching people in the vestry eating out of communal bowls of popcorn sitting on tables. People in the kitchen were using Coleman stoves to make this snack. The author knew that this community, like many rural areas relies on well water and septic systems. With the volume of water recently moving through this area, there would be no clean water available. With power out, there were few options to boil water, or anyplace sanitary for people to use restrooms. This community along with the two closest neighboring ones was isolated from normal movement. The population was heavily weighted toward elderly with existing health issues, functional and mobility issues.

A sheriff who as friends with the author from a neighboring county arrived (access from the home county was impossible). The Sheriff, fire chief, and author held a meeting where we agreed to form a Unified Area

Command with three levels of government, local, county, and state if we could get buy-in from the impacted county. The Sheriff worked his contacts, the author worked through state health, and the local county and state agreed. The author's direction from county delivered via state health was to do whatever we needed to do. Part of this effort included working requests through ESF-8 contacts at the state and federal level and back-notifying state EMA. State EMA was process driven and requests through that chain resulted in extensive delays.

Initial resource request was basic:

- Potable water (life safety issue)
- State Forestry Over-Head team to establish the ICS (organization is critical plus documentation)
- Strong backs to help locals and emergency transport (life-sustaining and stabilization)
- Medical supplies for community support due to restricted access to health care (life-sustaining and stabilization)

Over the course of the week the author lived and slept on the concrete floor of the fire chief' office a number of issues were dealt with.

- Burning of waste piling up from residential clean-up efforts
- Opening a clinic staffed by National Guard medical staff (this included the operation being shut down temporarily by one officer who did not want military medical staff treating civilians, even though that is exactly what the guard members did when in their civilian roles). The author had to reach back to find an officer who outranked the on-scene officer to get the clinic back open
- Emptying barns of wet hay that was threatening to combust and burn down the buildings.
- Building a lasting distaste for flavored water as that is what showed up on pallets via Chinook airlift as potable water.

When it was all said and done, the implementation of the ICS, including regular briefings and development of Incident Action Plans, was key in reimbursement for all resource requests and recovery efforts. For his efforts or making up a plan as he went, the author was awarded the Chiefs' Medal from State Health and a letter of commendation from the Governor.

Lessons learned

The Emergency Manager at some point of their career will find themselves faced with situations for which they have no immediate answer. Falling back on priorities for survivor well-being, relationships and common sense will have to be your guide. Don't try and eat an elephant in one sitting, take it a bite at a time and set clear objectives to make sure you're moving in the right direction. Diplomacy is preferred but when it comes to life safety issues, the Emergency Manager will need to make a choice, support the survivors or don't rock the boat. Rocking the boat can impact your career down the road so be comfortable whichever way you choose.

References

1. Waugh, W. L., Jr.; Tierney, K., Eds. *ICMA Green Book Emergency Management: Principles and Practices for Local Government*, 2nd ed. ICMA, 2009, p. 1.
2. Haddow, G. D.; Bullock, J. A.; Coppola, D. P. *Introduction to Emergency Management*, 6th ed.; Elsevier: Oxford; Cambridge, MA, 2017. p xiii, 1, 15, 16, 18, 19, 36, 40–46, 46–49, 52–57, 61–64, 66–68, 69–70, 73–74, 99–100, 205, 207, 418.
3. Gawande, A. *The Checklist Manifesto: How to Get Things Right; Metropolitan Books*; Henry Holt & Company: New York & London, 2009. p 5. 79.
4. Simon, H. A. *Administrative Behavior: A Study of Decision-Making Processes in Administrative Organizations*, 4th ed.; The Free Press: New York & London, 1997, p. 87.
5. Adamski, T.; Kline, B.; Tyrell, T. FEMA Reorganization and the Response to Hurricane Disaster Relief. *Perspect. Public Aff.* **2006**, *3*, 22–23.
6. Stallard, D.; Sanger, K. The Failure of Success Revisited; In *The Nathan Solution to the Bathsheba Syndrome*; 2018; Vol. *98*(4). https://www.academia.edu/6570487/Nathan_Solution_to_the_Bathsheba_Syndrome (Accessed 04.08.2018).
7. The Lightning Press. *About the Military Decision Making Process (MDMP)*, 2018. https://www.thelightningpress.com/about-the-military-decisionmaking-process-mdmp/ (Accessed 24.07).
8. The Lightning Press. *Military Decision Making Process (MDMP)*, 2018. https://www.the-lightningpress.com/military-decision-making-process-mdmp/ (Accessed 24.07.2018).
9. Rule, J. N. *A Symbiotic Relationship: The OODA Loop, Intuition, and Strategic Thought*; The Strategy Research Project, 2013.
10. Gladwell, M. M. *Blink: The Power of Thinking Without Thinking*; Little, Brown & Company: New York, 2005. p. 184.
11. McKay, B.; McKay, K. *The Tao of Boyd: How to Master the OODA Loop*. https://www.artofmanliness.com/articles/ooda-loop/.
12. Hammond, G. T. The Mind of War: John Boyd and American Security; In *Smithsonian Books*; 2004. Quoted in McKay, B.; McKay, K. The Tao of Boyd: How to Master the OODA Loop. https://www.artofmanliness.com/articles/ooda-loop/.
13. Osinga, F. P. *Science, Strategy and War: The Strategic Theory of John Boyd*; Taylor and Francis: New York, 2007.
14. Pearson, T. *The Ultimate Guide to the OODA Loop: How to Turn Uncertainty into Opportunity*, 2018. https://taylorpearson.me/ooda-loop/ (Accessed 05.04.2018).
15. Hightower, T. *Boyd's OODA Loop and How We Use It*; Tactical Response, 2018. https://www.tacticalresponse.com/blogs/library/18649427-boyd-s-o-o-d-a-loop-and-how-we-use-it> (Accessed 26.07.2018).
16. Lubitz, D. K.; Beakley, J. E.; Patricelli, F. *'All Hazards Approach' to Disaster Management: The Role of Information and Knowledge Management, Boyd's OODA Loop, and Network-Centricity*, 2018. https://www.ncbi.nlm.nih.gov/pubmed/18479475 (Accessed 07.07.2018) (PubMed).
17. Buchanan, L.; O'Connell, A. *A Brief History of Decision Making*; Harvard Business Review, 2018. https://hbr.org//2006/01/a-brief-history-of-decision-making. (Accessed 02.01.2018).
18. Stanford Encyclopedia of Philosophy *Aristotle's Ethics*, 2018. https://plato.stanford.edu/entries/aristotle-ethics/ (Accessed 21.03.2018).

19. Krueger, J. *Reason and Emotion: A Note on Plato, Darwin and Damasio*; Psychology Today, 2018. https://www.psychologytoday.com/us/blog/one-among-many/201006/reason-and-emotion-note-plato-darwin-and-damasio (Accessed 01.04.2018).
20. Kahneman, D. *Thinking, Fast and Slow*. Ferrar, Strauss and Giroux, 2011; p 8, 11, 12, 28, 80–81, 84–85, 88, 90, 131, 206, 207, 224–225, 324, 333, 324.
21. Mansbridge, J. J., Ed. *Beyond Self Interest*; Chicago & London: University Press, 1990.
22. Levy, J. M. *Contemporary Urban Planning*. 2nd ed.; Prentice Hall: New Jersey, 1998; p 274, 275, 278.
23. Braybrooke, D.; Lindblom, C. E. *A Strategy of Decision: Policy Evaluation as a Social Process*. Collier Macmillian Publishers: London, 1970; p vi, vii.
24. Lewis, M. *The Undoing Project*. W.W. Norton & Company, 2012; p 164, 334, 336.
25. Rogers, P.; Blenko, M. W. *Who Has the D? How Clear Decision Roles Enhance Organizational Performance*; Harvard Business Review, 2018. https://hbr.org/2006/01/who-has-the-d-how-clear-decision-roles-enhance-organizational-performance (Accessed 02.01.2018).
26. Kahneman, D.; Rosenfield, A. M.; Rosenfield, L. G.; Blaser, T. *Noise: How to Overcome the High, Hidden Costs of Inconsistent Decision Making*; Harvard Business Review, 2018. https://hbr.org/2016/10/noise (Accessed 19.02.2018).
27. Jacobs, A. *How to Think: A Survival Guide for a World at Odds*. Crown Publishing, A Division of Penguin Random House, LLC: New York, 2017; p 16, 17, 22, 58–63, 86, 87, 155, 156.
28. Haidt, J. *The Righteous Mind; Why Good People are Divided by Politics and Religion*. Vintage Books, A Division of Random House, Inc., 2012; p xx, xxi, 32.
29. Eliot, T. S. *The Sacred Wood: Essay on Poetry and Criticism*; Alfred A. Knopf: New York, 1921. Quoted in Jacobs, A. How to Think: A Survival Guide for a World at Odds; Crown Publishing, a Division of Penguin Random House, LLC: New York, 2017; p. 16, 17, 86, 87.
30. Wilson, T. *Strangers to Ourselves: Discovering the Adaptive Unconscious*; Harvard University Press: Cambridge, 2002, p. 12.
31. Thaler, R. H. *Misbehaving: The making of Behavioral Economics*; W.W. Norton & Company: New York; London, 2015; p 19.
32. Ovid; Johnston, I. (translator). *Media, in Metamorphosis, Book VII*. http://johnstoniatexts.x10host.com/ovid/ovid7html.html.
33. Buonamano, D. *Brain Bugs: How the Brain's Flaws Shape Our Lives*; W.W. Norton & Company: New York & London, 2011.2. 3.
34. Fine, C. A. *Mind of Its Own: How Your Brain Distorts and Deceives*; W.W. Norton & Company, 2006; pp 82–84.
35. Bornstein, R. F. Exposure and Affect: Overview and Meta-Analysis of Research, 1968–1987. *Psychol. Bull.* **1989**, *106*, 265–289. Quoted in Buonamano, D. Brain Bugs: How the Brain's Flaws Shape Our Lives; W.W. Norton & Company: New York & London; p. 2, 3.
36. Kuhn, D. Children and Adults as Intuitive Scientists. *Psychol. Rev.* **1989**, *96*, 764–789. Quoted in Haidt, J. The Righteous Mind; Why Good People are Divided by Politics and Religion; Vintage Books, A Division of Random House, Inc., 2012; p. xx, 32.
37. Janis, I. L. *Groupthink: Psychological Studies of Policy Decisions and Fiascos*, 2nd ed.; Houghton Mifflin Company: Boston, 1982; p vii, viii, 2, 5–9, 89–90.
38. Le Bon, G. *The Crowd*; The Viking Press: New York, 1964.
39. Rude, G. *The Crowd in History: A Study of Popular Disturbances in France And England, 1730-1848*; John Eiles & Sons: New York, 1964.
40. Kingdon, J. W. *Agendas, Alternatives and Public Policies*; Harper Collins: New York, 1995; pp 89–91.
41. Hoffer, E. *The True Believer: Thoughts on the Nature of Mass Movements*; Harper & Rowe: New York, 1951.
42. Sharot, T. *The Influential Mind: What the Brain Reveals About Our Power to Change Others*; Henry Holt and Company: New York, 2017; pp 15–17, 25.

43. Apenwarr. *The Curse of Smart People*, 2018. https://apenwarr.ca/log/20140701 (Accessed 29.03.2018). Quoted in Jacobs, A. How to Think: A Survival Guide for a World at Odds; Crown Publishing, A Division of Penguin Random House, LLC: New York, 2017; p. 58.
44. Kakkar, H.; Sivanathan, N. *Why We Prefer Dominant Leaders in Uncertain Times*; Harvard Business Review, 2017. https://hbr.org/2017/08/why-we-prefer-dominant-leaders-in-uncertain-times%20accessed%2001//20/2018/ (Accessed 11.08.2017).
45. Meehl, P. Causes and Effects of My Disturbing Little Book. *J. Pers. Assess.* **1986**, *50*, 370–375.
46. Finkelstein, S.; Whitehead, J.; Campbell, A. *Think Again: Why Good Leaders Make Bad Decisions and How to Keep It From Happening to You*; Harvard University Press: Boston, 2008; p x–xii, 8–11, 52, 76, 77, 114, 160–169.
47. Allison, G. T. Conceptual Models and the Cuban Missile Crisis. *Am. Polit. Sci. Rev.* **1969**, LXIII(3). Reproduced with permission in Classics of Public Policy; Pierson-Longman: New York, 248–271.
48. Niskanen, W. A., Jr. *Bureaucracy and Representative Government*; Aldine-Atherton: Chicago, 1971.5. 21–22, 38.
49. Niskanen, W. A., Jr. Bureaucrats and Politicians. *J. Law Econ.* **1975**, *18*, 617–643.
50. Global Centre for Public Service Excellence *What We Offer to UNDP Offices and Partner Governments*, 2018. http://www.undp.org/content/dam/undp/library/capacity-development/English/Singapore%20Centre/GCPSE_OurServices%203July17.pdf (Accessed 16.10.2018).
51. Wilson, J.; Bureaucracy, Q. *What Government Agencies Do and Why They Do It*; Basic Books: New York, 1998; p ix, xv, 199.
52. Golden, M. M. *What Motivates Bureaucracy? Politics and Administration During the Reagan Years*; Columbia University Press: New York, 2000; p 1, 8–26, 72, 171.
53. Peters, G., Pierre, J., Eds. *The Handbook of Public Administration*; Sage Publication: London; New Delhi; Singapore, 2007; p 200.
54. Moe, T. M. Regulatory Performance and Presidential Administration. *Am. J. Polit. Sci.* **1982**, *26*, 197–223.
55. Wood, B. D.; Waterman, R. W. The Dynamics of Political Control of the Bureaucracy. *Am. Polit. Sci. Rev.* **1991**, *85*(3), 801–828.
56. Lipsky, M. *Street Level Bureaucrats as Policy Makers From Street Level Bureaucracy: Dilemmas of the Individual in Public Service*; Russel Sage Foundation, 1980. Reprinted with permission in Classics of Public Policy; Pierson-Longman: New York, 2005; p. 2, 51.
57. Wildavsky, A.; Pressman, J. L. *Implementation*; 2nd ed.; The University of California: Berkeley; Los Angeles, 1973.
58. Boorman, D. J. *Reducing Flight Crew Errors and Minimizing New Error Modes with Electronic Checklists*. In *Proceedings of the international Conference on Human Computer Interactions in Aeronautics*; 2000; pp 457–463. (Editions Ce'Paude's). Quoted in Gawande, A. The Checklist Manifesto: How to Get Things Right; Metropolitan Books, Henry Holt & Company: New York & London, 2009; p. 122–124.
59. Boorman, D. J. Today's Electronic Checklists Reduce Likelihood of Crew Errors and Help Prevent Mishaps. *ICAO J.* **2001**, *56*, 17–20. Quoted in Gawande, A. The Checklist Manifesto: How to Get Things Right; Metropolitan Books, Henry Holt & Company: New York & London, 2009; p. 122–124.
60. Wilson, E. O. *Sociobiology*; The Harvard University Press: Cambridge, 1975. 2980 , ix, xv, 199.
61. Hamilton, W. D. The Evolution of Altruistic Behavior. *Am. Nat.* **1963**, *97*, 354.
62. Axelrod, R. *The Evolution of Cooperation*; revised ed.; Basic Books: New York, 1984; pp. 20–23, 38, 39.
63. Drabek, T. E., Hoetmer, G. J., Eds. *Emergency Management: Principles and Practices for Local Government, The Municipal Management Series*; The International City Managers Association: Washington, DC, 1992, pp. 134, 135.
64. Gilbert, E. Crash Course. *The New York Times* **2018**.

65. Schwartz, J. Humans Are Making Hurricanes Worse. Here's How; *The New York Times* **2018**, https://www.nytimes.com/2018/09/19/climate/humans-hurricanes-causes-effects.html.

66. Teutsch, K. *Effective Disaster Management Strategies in the 21st Century*; Government Technology, 2018. https://www.govtech.com/em/disaster/Effective-Disaster-Management-Strategies.html (Accessed 11.12.2018).

67. Waugh, W. L.; Tierney, K. *Emergency Management: Principles and Practice for Local Government*. 2nd ed.; ICMA: Washington, DC, 2014; p 4, 6, 9, 15, 18, 98–100, 260, citing Ref. 69.

68. Waugh, W. L., Jr. *Terrorism as Disaster*; Disaster Research, 2017.

69. Waugh, W. L., Jr. Mechanisms for Collaborations: ICS, NIMS, and the Problem of Command and Control. In *The Collaborative Manager*; Georgetown University Press, 2008.

70. MacDonald, H. Why the FBI Didn't Stop 9/11: Blame Elite Beliefs and Clinton-Era Edicts; *City J.* **2018**. https://www.city-journal.org/html/why-fbi-didn%E2%80%99t-stop-911-12373.html (Accessed 06.12.2018).

71. Eichwald, K. The Deafness Before the Storm: It Was Perhaps the Most Famous Presidential Briefing in History. *The New York Times* **2012**.

72. *National Commission on Terrorist Attacks Upon the United States. Foresight and Hindsight*, 2018; https://govinfo.library.unt.edu/911/report/911Report_Ch11.htm (Accessed 27.11.2018).

73 Government Executive. *Katrina 10: An Oral History*. https://www.govexec.com/feature/katrina-10/.

74. *Mass Medical Evacuation: Hurricane Katrina and Nursing Experiences at the New Orleans Airport*; U.S. National Library of Medicine, 2018. https://www.ncbi.nlm.nih.gov/pubmed/17517364 (Accessed 26.07.2018).

75. Bluestein, G. *Army General Recalls Katrina Aftermath*; The Associated Press, 2018. http://www.washingtonpost.com/wp-dyn/content/article/2006/09/07/AR2006090700163_pf.html (Accessed 02.08.2018).

76. The Washington Post. *Want Better, Smaller Governments? Hire Another Million Federal Bureaucrats*, 2018. https://www.washingtonpost.com/opinions/want-better-smaller-government-hire-1-million-more-federal-bureaucrats/2014/08/29/c0bc1480-2c72-11e4-994d-202962a9150c_story.html?noredirect=on&utm_term=.6e290fb1796a (Accessed 12.11.2018).

77. SSRN. *Public Accountability: Performance Measurement, the Extended State, and the Search for Trust*, 2011. https://papers.ssrn.com/sol3/papers.cfm?abstract_id=1875024.

78. The Economist. *Weather-Related Disasters Are Increasing: But the Number of Deaths Caused by Them Is Falling*, 2018. https://www.economist.com/graphic-detail/2017/08/29/weather-related-disasters-are-increasing (Accessed 01.08.2018).

79. Crosset, K. M.; et al. Population Trends Along the Coastal United States: 1980–2008. Coastal Trends Report Series; 2004. Quoted in Waugh, W. L.; Tierney, K. Emergency Management: Principles and Practice for Local Government, 2nd ed.; An ICMA Green Book, ICMA.

80. Sutter, J. D.; Santiago, L. *Hurricane Maria Death Toll May Be More Than 4,600 in Puerto Rico*; CNN, 2018. https://www.cnn.com/2018/05/29/us/puerto-rico-hurricane-maria-death-toll/index.html (Accessed 01.08.2018).

81. Fuller, T.; Turkewitz, J. Dry Weather and Roaring Wildfires: 'New Normal.'. *The New York Times* **2018**, A13.

82. Grieving, S.; Fleischhauer, M.; Wanczura, S. Management of Natural Hazards in Europe: The Role of Spatial Planning in Selected EU Member States. *J. Environ. Plan. Manage.* **2006**, *49*(5), 739–757.

83. Waugh, W. L. Mechanisms for Collaboration in Emergency Management, ICS, NIMS, and the Problem of Command and Control. In *The Collaborative Public Manager: New Ideas for the 21st Century*, 2009. Quoted in Waugh, W. L.; Tierney, K. Emergency Management: Principles and Practice for Local Government, 2nd ed.; An ICMA Green Book, ICMA.

84. Drabek, T. E. *The Professional Emergency Manager*; Institute for Behavioral Science: Boulder, 1987.

85. Waugh, W. L.; Tierney, K. Emergency Management: Principles and Practice for Local Government, 2nd ed.; *An ICMA Green Book*; ICMA, 2007, pp. 7–9.

86. Government Technology. *Using Data Analytics as a Viable Way to Facilitate Resilience and Better Recovery*, 2018. https://www.govtech.com/em/preparedness/Using-Data-Analytics-as-a-Viable-Way-to-Facilitate-Resilience-and-Better-Recovery.html (Accessed 11.08.2018).

87. McGuire, M. The New Professionalization and Collaborative Activity in Local Emergency management. In *The Cooperative Public Manager…*, 2009, p 73. 47.

88. Rackley, K.; Standard, A. *New App Helps People Stay Safe During Natural Disasters*; Government Technology, 2018. https://www.govtech.com/em/preparedness/New-App-Helps-People-Stay-Safe-During-Natural-Disasters.html (Accessed 16.07.2018).

89. NOAA. *National Centers for Environmental Information. National Climate Report—March 2019*, 2019. https://www.ncdc.noaa.gov/sotc/national/?Set-Language=arte (Accessed 15.09.2018).

90. Briscoe, T. Climate Report a Warning for Midwest. *The Chicago Tribune* **2018**, *1*, 15.

91. Robles, F.; Patel, J. K. With Paltry Housing Aid, Lives in Puerto Rico Remain in Ruins. *The New York Times* **2018**, A13–A16.

92. McKay, J. *Long-Term Recovery Never Ends for Some After Natural Disasters*; Government Technology, 2018. https://www.govtech.com/em/disaster/Long-Term-Recovery-Never-Ends-for-Some-After-Natural-Disasters.html (Accessed 11.12.2018).

93. Government Technology. *Survivors of Disaster Often Victims a Second Time*, 2018. https://www.govtech.com/em/disaster/Survivors-of-Disaster-Often-Victims-a-Second-Time.html (Accessed 12.06.2018).

94. Schwartz, S. *A Government Technology Case Study*; AT&T, 2018.

95. Salamon, L.; et al. *Nonprofit Employment Bulletin: No. 39*. The Johns Hopkins Nonprofit Economic Data Project, 2012; p 2. Quoted in DiIulio, J. Bring Back the Bureaucrats, 2014: Why More Federal Bureaucrats Will Lead to Better (and Smaller) Government. Templeton Press: Conshohocken, 2014.

96. Homeland Security. *Critical Infrastructure Sectors*. https://www.dhs.gov/cisa/critical-infrastructure-sectors.

97. Fagel, M. J. *Fagel's Principles of Emergency Management and Emergency Operations Centers (EOC)*; CRC Press, 2011.

98. Henderson, D. A.; Inglesby, T. V.; O'Toole, T. *Bioterrorism: Guidelines for Medical and Public Health Management*; JAMA, 2002.

99. Glarum, J.; Birou, D.; Cetaruk, E. *Hospital Response Teams*; Elsevier, 2010.

100. United Nations Department of Economic and Social Affairs, *Majority of the World's Cities Highly Exposed to Disasters, UN DESA Warns on World Cities Day*, 2018. https://www.un.org/development/desa/en/news/population/world-cities-day-2018.html (Accessed 18.09.2018).

101. Schwartz, J. The Cities of Tomorrow. *The New York Times* **2018**, F1–F8.

102. Keim, M.; Floods, E. *Koenig and Schultz's Disaster Medicine*, 2nd ed.; Cambridge University Press, 2016; pp 602–616.

103. National Weather Service. *Flood Safety Tips and Resources*, 2018. https://www.weather.gov/safety/flood (Accessed 22.07.2018).

104. News and Observer. *Florence Bathed NC in Raw Sewage. New Figures Show It Was Even Worse Than We Thought*, 2019. https://www.newsobserver.com/news/business/article223328915.html (Accessed 22.02.2019).

105. Environment. *The EU Floods Directive*; European Commission, 2019. http://ec.europa.eu/environment/water/flood_risk/com.htm (Accessed 11.04.2019).

106. Schwarze, R.; Schwindt, M.; et al. *Natural Hazard Insurance in Europe: Tailored Responses to Climate Change Are Needed*; Wiley Online Library, 2010. https://www.researchgate.net/publication/227523973_Natural_Hazard_Insurance_in_Europe_Tailored_Responses_to_Climate_Change_are_Needed>.

107. Government Technology. *Houston Must Build Resilience to Avoid Next Flood, Hurricane Disaster, Insurer's Report Finds*, 2018. https://www.govtech.com/em/preparedness/Houston-Must-Build-Resilience-to-Avoid-Next-Flood-Hurricane-Disaster-Insurers-Report-Finds.html (Accessed 10.07.2018).

108. Miami Herald. *Your Coastal Property Has Already Lost Value to Sea Rise. This Site Can Tell You How Much*, 2018. https://www.miamiherald.com/news/local/environment/article215421425.html (Accessed 01.08.2018).

109. Miller, J. A. *Credit Downgraded Threat as a Non-regulatory Driver for Flood Risk Mitigation and Sea Level Rise Adaptation*; University of Pennsylvania's Scholarly Commons, 2018. Quoted by Weinstein, S.; et al. disaster_law@lists.berkeley.edu.

110. Penn Libraries Scholarly Commons. *Credit Downgrade Threat as a Non-Regulatory Driver for Flood Risk Mitigation and Sea Level Rise Adaptation*, 2019. https://repository.upenn.edu/mes_capstones/73/ (Accessed 24.01.2019).

111. Houston Chronicle. *Flood Games*, 2019. https://repository.upenn.edu/mes_capstones/73/ (Accessed 18.03.2019).

112. Barron, J. City Soaking Up Lessons From Denmark on Floods. *The New York Times* **2018**, A22.

113. Stratton, J. W. *Earthquakes, in The Public Health Consequences of Disasters: 1989, A CDC Monograph*; Department of Health and Human Services, U.S. Public Health Service, Centers for Disease Control and Prevention, 1989; p 13.

114. Mid-America Earthquake Center. *New Madrid Seismic Zone Catastrophic Earthquake Response Planning Project*, 2018. http://cusec.org/documents/scenarios/2009_Scenario_MAE_Center_Vol_I.pdf (Accessed 09.07.2018).

115. Government Technology. *Catastrophic Earthquakes Could Leave 250,000–400,000 Refugees in California*, 2018. https://www.govtech.com/em/preparedness/Catastrophic-Earthquakes-Could-Leave-250000-400000-Refugees-in-California.html (Accessed 30.10.2018).

116. FEMA. *Building Codes in the New Madrid Seismic Zone (NMSZ)*, 2018. https://www.iccsafe.org/gr/Documents/AdoptionToolkit/nmsz_building_code_adoption.pdf (Accessed 12.07.2018).

117. The New York Times. *San Francisco's Big Seismic Gamble*, 2018. https://www.nytimes.com/interactive/2018/04/17/us/san-francisco-earthquake-seismic-gamble.html?mtrref=undefined&gwh=1E7EFA9CF5D3F253AB8C8B9EF78E0C69&gwt=pay (Accessed 21.09.2018).

118. Strategy+Business. *The Best Management Is Less Management*, 2019. https://www.strategy-business.com/article/The-Best-Management-Is-Less-Management (Accessed 05.01.2019).

119. The National Geographic. *These Are the Most Dangerous U.S. Volcanoes, Scientists Say*, 2018. https://www.nationalgeographic.com/science/2018/10/news-most-dangerous-volcanoes-usgs-list-geology.html.

120. Jones, L. *The Big Ones: How Natural Disasters Have Shaped Us and What We Can Do About Them*; Vols. 1–6; DoubleDay: New York, 2018; pp. 22, 52–56.

121. Baxter, P.; Volcanoes, J. *Koenig and Schultz's Disaster Medicine*, 2nd ed.; Cambridge University Press, 2016; pp 732–736.

122. Stratton, S. L. Tsunamis. In *Koenig and Schultz's Disaster Medicine: Comprehensive Principles and Practices*, 2nd ed; Cambridge University Press, 2016; p. 661–668; Ref. 2, pp. 52–54.

123. Science on a Sphere. *Tsunami Historical Series: Samoa—2009*, 2019. https://sos.noaa.gov/datasets/tsunami-historical-series-samoa-2009/ (Accessed 24.03.2019).

124. NBC. *List: Historic Tsunamis on California's Coast*, 2019. https://www.nbclosangeles.com/news/local/Earthquake-Tsunami-California-Waves-History-Damage-470704043.html (Accessed 24.03.2019).

125. Solomon, B. C. Hope All But Gone, Keeping Vigil for Missing in the Indonesian Disaster. *The New York Times* **2018**, A5.

126. Woods Hole Oceanographic, 2019. https://www.whoi.edu/ (Accessed 26.03.2019).

127. The New York Times. *Satellite Photos of Japan, Before and After the Quake and Tsunami*, 2019. http://archive.nytimes.com/www.nytimes.com/interactive/2011/03/13/world/asia/satellite-photos-japan-before-and-after-tsunami.html?_r=0 (Accessed 26.03.2019).

128. Science on a Sphere. *What Is Science on a Sphere®?*, 2019. https://sos.noaa.gov/What_is_SOS/ (Accessed 27.03.2019).

129. Springer Link. *Observations and Impacts From the 2010 Chilean and 2011 Japanese Tsunamis in California (USA)*, 2019. https://link.springer.com/article/10.1007/s00024-012-0527-z (Accessed 27.03.2019).

130. Ready.Gov. *Voluntary Organizations Active in Disaster*, 2019. https://www.ready.gov/voluntary-organizations-active-disaster (Accessed 15.03.2019).

131. Kidney Community Emergency Response. *Healthcare Coalition List by State*, 2019. https://www.kcercoalition.com/en/resources/professional-resources/cms-emergency-preparedness-rule/state-healthcare-coalitions-list/healthcare-coalition-list-by-state/ (Accessed 15.03.2019).

132. Government Technology. *The World Has Never Seen a Category 6 Hurricane, But the Day May Be Coming*, 2018. https://www.govtech.com/em/preparedness/The-World-Has-Never-Seen-a-Category-6-Hurricane-but-the-Day-May-be-Coming.html.

133. Government Technology. *Bigger, Wetter, Costlier: Studies Suggest Trend of Slower, Wetter Hurricanes as County Reviews Lessons Learned*, 2018. https://www.govtech.com/em/preparedness/Bigger-Wetter-Costlier-Studies-suggest-trend-of-slower-wetter-hurricanes-as-county-reviews-lessons-learned.html (Accessed 10.07.2018).

134. Government Technology. *Hurricane Michael Reminds Us It's Past Time to Get Smarter About Where We Build*, 2018. https://www.govtech.com/em/preparedness/Hurricane-Michael-Reminds-us-its-Past-Time-to-get-Smarter-About-Where-we-Build.html?utm_term=READ%20MORE&utm_campaign=It%27s%20Past%20Time%20to%20be%20Smarter%20About%20Where%20We%20Build&utm_content=email&utm_source=Act-On+Software&utm_medium=email (Accessed 26.10.2018).

135. Pierre-Louis, K. 3 Hurricane Misconceptions Scientists Want to Set Straight. *The New York Times* **2018**, A18.

136. 2017 Hurricane Season FEMA After-Action Report; 2018 (Accessed 12.07.2018).

137. U.S. Department of Health and Human Services. *The Emergency Prescription Assistance Program*, 2019. https://www.phe.gov/Preparedness/planning/epap/Pages/default.aspx (Accessed 02.02.2019).

138. ASPR Blog. *Modernizing the National Disaster Medical System to Meet the Health Security Threats of the 21st Century*, 2018. https://www.phe.gov/ASPRBlog/Lists/Posts/Post.aspx?ID=316.

139. Radio Farda. *Analysis: Iran's Failed Response to Natural Disasters*, 2018. https://en.radiofarda.com/a/iran-response-to-natural-disasters-weak/28864340.html (Accessed 11.12.2018).

140. Wightman, J. M.; Dice, W. H. Winter Storms and Hazards. In *Koenig and Schutlz's Disaster Medicine*, 2nd ed; Cambridge University Press, 2016, p 670.

141. Sanderson, L. M.; Gregg, M. B. Tornados. In *The Public Health Consequences of Disasters*, US Department of Health and Human Services, Public Health Service, 1989, p 39.

142. Patton, A. J. *Surviving the Storm: Sheltering in the May 2003 Tornadoes, Moore, Oklahoma*; Quick Response Report 163, Natural Hazards Research and Applications Center, University of Colorado: Boulder, 2003.

143. Government Technology. *New App Helps People Stay Safe During Natural Disasters*, 2018. https://www.govtech.com/em/preparedness/New-App-Helps-People-Stay-Safe-During-Natural-Disasters.html.

144. Pyne, S. J. *Fire Fundamentals: A Primer on Wildland Fire for Journalists*, 2018. https://en.radiofarda.com/a/iran-response-to-natural-disasters-weak/28864340.html (Accessed 2018).

145. AMS100. *Free Access Water, Drought, Climate Change, and Conflict in Syria*, 2018. https://journals.ametsoc.org/doi/full/10.1175/WCAS-D-13-00059.1 (Accessed 12.03.2018).

146. ReliefWeb. *Editor's Pick: 10 Violent Water Conflicts*, 2018. https://reliefweb.int/report/world/editor-s-pick-10-violent-water-conflicts (Accessed 21.11.2018).

147. National Integrated Drought Information Systems. *Advancing Drought Science and Preparedness across the Nation*, 2018. https://www.drought.gov/drought/ (Accessed 21.08.2018).

148. No-Till Farmer. *Possible Solution for Helping Ogallala Aquifer Woes?*, 2018. https://www.no-tillfarmer.com/blogs/1-covering-no-till/post/6097-possible-solution-for-helping-ogallala-aquifer-woes (Accessed 28.11.2018).

149. Schilling, M. K. Dust to Dust. *Newsweek* **2018**, 32.

150. Government Technology. *Increasingly Intense Wildfires Becoming More of a Threat to Water Supplies*, 2018. https://www.govtech.com/em/preparedness/Increasingly-Intense-Wildfires-Becoming-More-of-a-Threat-to-Water-Supplies.html.

151. Anderson, C.; Cowell, A. A Very Strange Year in Sweden, as Wildfires Reach the Arctic Circle. *The New York Times* **2018**, 15.

152. NCA 2018. *Fourth National Climate Assessment*, 2018. https://nca2018.globalchange.gov/ (Accessed 03.03.2018).

153. Holden, C. How to Live With Wildfires. *The New York Times* **2018**, 7.

154. Serna, J. San Francisco Sets All-Time Heat Record Downtown at 106 Degrees During State's Hottest Recorded Summer. *Los Angeles Times* **2017**.

155. United States Environmental Protection Agency. *The Excessive Heat Events Guidebook*, 2018. https://www.epa.gov/heat-islands (Accessed 22.09.2018).

156. Adrianopoli, C.; Jacoby, I. *Koenig and Schultz's Disaster Medicine* 2nd ed.; Cambridge University Press, 2016; p 699.

157. Why Do Older Patients Die in a Heatwave? *QJM* **2005** https://academic.oup.com/qjmed/article/98/3/227/1538832.

158. Sengupta, S. In India, Summer Heat Becomes a 'Silent Killer': 111 Days Hit Poor the Hardest. *The New York Times* **2018**, A1. A9.

159. Part II, Technological Hazards. In *Principal Threats Facing Communities and Local Emergency Management Coordinators*, FEMA, 1993, pp 249–251.

160. National Center for Biotechnology Information. *Three Mile Island and Bhopal: Lessons Learned and Not Learned*, 2018. https://www.ncbi.nlm.nih.gov/books/NBK217577/ (Accessed 22.07.2018).

161. Kushner, H. W., Ed *Terrorism in the 21st Century*, Sage Publications, 2001. 44(6).

162. Bolz, F., Jr.; Dudonis, K. J.; Schultz, D. P. *The Counterterrorism Handbook: Tactics, Procedure, and Techniques*, 4th ed.; CRC Press: Boca Raton; London; New York, 2009.

163. Dickey, C. *Securing the City: Inside America's Best Counter Force—The NYPD*; Simon and Schuster: New York; London, 2010.

164. Priest, R. *Hot Zone: A Terrifying True Story*; Anchor Books: New York, 1994.

165. FEMA: In or Out; Department of Homeland Security, Office of the Inspector General Report, 2009. Quoted in Haddow, G. D.; Bullock, J. A.; Coppola, D. P. Introduction to Emergency Management, 6th ed.; Elsevier: Oxford and Cambridge, MA, 2017; p. 418, 419.

166. Nixon, R.; Hirschfeld Davis, J.; Glanz, J. Storm Present Test for FEMA: Preparations Driven by Lessons From 2017. *The New York Times* **2018**, 1, 17.

167. GAO: Report to Congresses Addresses. In *2017 Hurricanes and Wildfires—Initial Observations on the Federal Response and Key Recovery Challenges*, 2018, pp 18–472. GAO Report.

168. U.S. Government Accountability Office. *2017 Hurricanes and Wildfires: Initial Observations on the Federal Response and Key Recovery Challenges*, 2018. https://www.gao.gov/products/GAO-18-472 (Accessed 06.09.2018).

169. Countering False Information on Social Media in Disasters and Emergencies: Social Media Working Group for Emergency Services and Disaster Management; US Department of Homeland Security, Science and Technology, 2018.

170. Knowles, G. (Op-Ed.). *The New York Times* **2018**, A29.

171. FEMA. *Report on Costs and Benefits of Natural Hazard Mitigation*, 2018. https://www.fema.gov/media-library/assets/documents/3459 (Accessed 29.10.2018).

172. National Disaster Legal Aid. *The Storm After the Storm*, 2019. https://www.disasterlegalaid.org/news/article.700052-The_Storm_After_the_Storm (Accessed 07.02.2019).

173. The Atlantic. *The Crisis at Puerto Rico's Hospitals*, 2017. https://www.theatlantic.com/health/archive/2017/09/the-crisis-at-puerto-ricos-hospitals/541131/.

174. U.S. Legal News. *Battered Puerto Rico Hospitals on Life Support After Hurricane Maria*, 2019. https://www.reuters.com/article/legal-us-storm-maria-puertorico-hospital/battered-puerto-rico-hospitals-on-life-support-after-hurricane-maria-idUSKCN1C11DA (Accessed 20.02.2019).

175. Kishore, N.; Marquez, D.; et al. Mortality in Puerto Rico After Hurricane Maria. 2017 Hurricane Season FEMA After-Action Report. *New Engl. Med. J.* **2018**, *379*, 162–170.

176. KFF. *The Recovery of Community Health Centers in Puerto Rico and the US Virgin Islands One Year after Hurricanes Maria and Irma*, 2018. https://www.kff.org/medicaid/issue-brief/the-recovery-of-community-health-centers-in-puerto-rico-and-the-us-virgin-islands-one-year-after-hurricanes-maria-and-irma/ (Accessed 16.03.2018).

177. Fernandez, M.; Panich-Linsman, I. 1 Year Later: Relief Stalls for Poorest in Houston. *The New York Times* **2018**, A9.

178. Wikipedia. *List of Cognitive Biases.* https://en.wikipedia.org/wiki/List_of_cognitive_biases.

179. Los Angeles Times. *Top FEMA Jobs: No Experience Required*, 2018. https://www.latimes.com/archives/la-xpm-2005-sep-09-na-fema9-story.html (Accessed 24.08.2018).

180. Los Angeles Times. *Put to Katrina's Test*, 2018. https://www.latimes.com/archives/la-xpm-2005-sep-11-na-plan11-story.html (Accessed 13.02.2018).

181. Gaouette, N.; Miller, A. C.; Mazzetti, M.; McManus, D.; Meyer, J.; Sack, K. Put to Katrina's Test. *The Los Angeles Times* **2005**.

182. Fox, T. Advice for Dealing with New Political Appointees in your Agency. *The Washington Post* **2013**.

183. Krumholz, N.; Forester, J. *To Be Professionally Articulate, Be Politically Effective. Making Equity Planning Work*; Temple University Press, 1990. Quoted in Stein, J.; Classic Readings in Urban Planning: An Introduction; McGraw-Hill: New York and St. Lewis, 1995.

184. Government Technology. *Hurricanes Florence, Michael Raise the Issue of Public Health During Disaster*, 2018. https://www.govtech.com/em/preparedness/Hurricanes-Florence-Michael-Raise-the-Issue-of-Public-Health-During-Disaster.html?utm_term=Hurricanes%20Florence%2C%20Michael%20Raise%20the%20Issue%20of%20Public%20Health%20During%20Disaster&utm_campaign=Hurricanes%20Raise%20the%20Issue%20of%20Public%20Health%20During%20Disasters&utm_content=email&utm_source=Act-On+Software&utm_medium=email (Accessed 12.11.2018).

185. Government Technology. *First Responders Are Beginning to Address Their Own Health*, 2018. https://www.govtech.com/em/preparedness/First-Responders-Are-Beginning-to-Address-Their-Own-Health-.html (Accessed 05.10.2018).

186. Psychology Today. *Psychopathy*. https://www.psychologytoday.com/us/basics/psychopathy.

187. The Telegraph. *How to Spot a Psychopath*. https://www.telegraph.co.uk/books/non-fiction/spot-psychopath/.

188. U.S. Government Accountability Office. *Federal Disaster Assistance: Federal Departments and Agencies Obligated at Least $277.6 Billion during Fiscal Years 2005 Through 2014*, 2018. https://www.gao.gov/products/GAO-16-797 (Accessed 22.11.2018).

189. PEW Trusts. *Federal Disaster Assistance Goes Beyond FEMA*, 2018. https://www.pewtrusts.org/en/research-and-analysis/fact-sheets/2017/09/federal-disaster-assistance-goes-beyond-fema (Accessed 22.11.2018).

190. DiIulio, J. *Bring Back the Bureaucrats: Why More Federal Workers Will Lead to Better (and Smaller!) Government. New Threats to Freedom*; Templeton Press: Coonshohocken, 2014, p. 35 citing Ref. 95, p. 2.

191. DiIulio, J. *Bring Back the Bureaucrats, 2014: Why More Federal Bureaucrats Will Lead to Better (and Smaller) Government*, Templeton Press: Conshohocken, 2014; pp 13–16. 39, 41, 42, 59–61, 129, 139, 141–142.

192. Acting Responsibly? Federal Contractors Frequently Put Workers' Lives and Livelihoods at Risk. Majority Committee Staff Report; U.S. Senate, Health, Education, Labor, and Pension Committee, 2013. Quoted in DiIulio, J. Bring Back the Bureaucrats, 2014: Why More Federal Bureaucrats Will Lead to Better (and Smaller) Government. Templeton Press: Conshohocken, 2014.

193. Interagency Contracting Agency Lessons Address Key management Challenges, But Additional Steps needed to Ensure Consistent Implementation of Policy Changes; U.S. Government Accountability Office, 2013. Quoted in DiIulio, J. Bring Back the Bureaucrats, 2014: Why More Federal Bureaucrats Will Lead to Better (and Smaller) Government. Templeton Press: Conshohocken, 2014.

194. Additional Steps Needed to Help Determine the Right Size and Composition of DOD's Total Workforce, and Defence Contracting Actions Needed to Increase Competition; U.S. Government Accountability Office, 2013. Quoted in DiIulio, J. Bring Back the Bureaucrats, 2014: Why More Federal Bureaucrats Will Lead to Better (and Smaller) Government. Templeton Press: Conshohocken, 2014.

195. Pettijohn, S. L. *The Nonprofit Sector in Brief: Public Charities, Giving, and Volunteering*; Urban Institute: Washington, DC, 2013; p 2. Quoted in DiIulio, J. Bring Back the Bureaucrats, 2014: Why More Federal Bureaucrats Will Lead to Better (and Smaller) Government. Templeton Press: Conshohocken, 2014.

196. Sharkansky, I. Policy Making and Service Delivery on the Margins of Government: The Case of Contractors. *Public Adm. Rev.* 1980, 40(2), 116–123.

197. Pogo. *Contractors and the True Size of Government*, 2018. https://www.pogo.org/analysis/2017/10/contractors-and-true-size-of-government/ (Accessed 13.11.2018).

198. Cambridge Core. *The 2015 John Gaus Award Lecture: Vision + Action = Faithful Execution: Why Government Daydreams and How to Stop the Cascade of Breakdowns That Now Haunts It*; 2018. https://www.cambridge.org/core/journals/ps-political-science-and-politics/article/2015-john-gaus-award-lecture-vision-action-faithful-execution-why-government-daydreams-and-how-to-stop-the-cascade-of-breakdowns-that-now-haunts-it/8E843F-CEAB511ADC29B57B2CE45DDAF5# (Accessed 04.11.2018).

199. Allison, B. Good Enough for Government Work. *The Washington Post* **2013**. Quoted in DiIulio, J. Bring Back the Bureaucrats, 2014: Why More Federal Bureaucrats Will Lead to Better (and Smaller) Government. Templeton Press: Conshohocken, 2014.

200. The Pew Charitable Trust. *What We Don't Know About State Spending on Natural Disasters Could Cost Us: Data Limitations, Their Implications for Policy-Making, and Strategies for Improvement*, 2018. https://www.pewtrusts.org/en/research-and-analysis/issue-briefs/2018/09/natural-disaster-mitigation-spending--not-comprehensively-tracked.

201. National Institute of Building Sciences. *Natural Hazard Mitigation Saves: 2017 Interim Report*, 2018. https://www.nibs.org/page/mitigation/saves Quoted in The Pew Charitable Trust. What We Don't Know about State Spending on Natural Disasters Could Cost Us: Data Limitations, Their Implications for Policy-Making, and Strategies for Improvement. https://www.pewtrusts.org/en/research-and-analysis/issue-briefs/2018/09/natural-disaster-mitigation-spending--not-comprehensively-tracked.

202. The Pew Charitable Trust. *Natural Disaster Mitigation Spending Not Comprehensively Tracked*, 2018. https://www.pewtrusts.org/en/research-and-analysis/issue-briefs/2018/09/natural-disaster-mitigation-spending--not-comprehensively-tracked.

203. Government Executive. *New Federal Senior Executives Are Being Left in the Dark as They Start Their Jobs*, 2018. https://www.govexec.com/management/2018/02/new-federal-senior-executives-left-in-the-dark/146063/ (Accessed 16.11.2018).

204. FEDWeek. *OPM Reports on Why SES Members Leave Government*, 2018. https://www.fedweek.com/issue-briefs/opm-reports-ses-members-leave-government/ (Accessed 16.11.2018).

205. Knowles, S. G. *The Disaster Experts: Mastering Risk in Modern America*; The University of Pennsylvania Press: Philadelphia, 2011; p 255.
206. Lewis, M. *The Fifth Risk*, W.W. Norton & Company: New York; London, 2018; pp 170–171.
207. Glazer, N. The Attack on the Professions. *Commentary* **1978**, *66*(5), 34.
208. Climate Wire. *Miami Beach Developer Dismisses Rising Waters as 'Paranoia'*, 2019. https://www.ee-news.net/climatewire/2019/01/28/stories/1060118683?show_login=1&t=https%3A%2F%2F-www.eenews.net%2Fclimatewire%2F2019%2F01%2F28%2Fstories%2F1060118683.
209. Columbia Law School: Sabin Center for Climate Change Law. *Breaking the Cycle of "Flood-Build-Repeat": Local and State Options to Improve Substantial Damage and Improvement Standards in the National Flood Insurance Program*, 2019. http://columbiaclimatelaw.com/files/2019/01/Adler-Scata-2019-01-Breaking-the-Cycle-Final.pdf.
210. Barron, J. *Thinking and Deciding*, 4th ed.; Cambridge University Press: New York, 2009.
211. Klein, G. A. *Sources of Power: How People Make Decisions*; MIT Press: Cambridge, 1998.
212. Tversky, A.; Kahneman, D. Extensional Versus Intuitive Reasoning: The Conjunction Fallacy in Probability Judgement. *Psychol. Rev.* **1983**, *90*, 293–315.
213. Tetlock, P. *Expert Political Judgement: How Good Is It? How Can We Know?* New ed.; Princeton University Press: Princeton; Oxford, 2017.
214. Haidt, J. Quoted by Nichols, T.; The Death of Expertise: The Campaign Against Established Knowledge and Why It Matters; Oxford University Press: New York, 2017.
215. Brooks, D. Students Learn From People They Love: Putting Relationship Quality at the Center of Education. *The New York Times* **2019**, A23.
216. The Washington Post. *The Science of Protecting People's Feelings: Why We Pretend All Opinions Are Equal*, 2019. https://www.washingtonpost.com/news/energy-environ-ment/wp/2015/03/10/the-science-of-protecting-peoples-feelings-why-we-pretend-all-opinions-are-equal/?utm_term=.8cf08d8c1062 (Accessed 21.01.2019).
217. PEW Trust *When Disaster Strikes, Governments Put All Hands on Deck*, 2018. https://www.pewtrusts.org/en/research-and-analysis/articles/2017/04/17/when-disaster-strikes-governments-put-all-hands-on-deck (Accessed 19.10.2018).
218. Nichols, T. *The Death of Expertise: The Campaign Against Established Knowledge and Why It Matters*; Oxford University Press: New York, 2017.
219. Fink, S. A. Tech Answer to Disaster Aid is Falling Short. *The New York Times August 11*, **2019**; pp. 1, 14, 15.
220. Keller, R. S. Keeping Disaster Human: Empathy, Systemization, and the Law. *Minn. J. Law Sci. Technol.* 2016, *17*(1), 1.
221. Hsu, S. S.; Glasser, S. B. FEMA. Director Singled Out by Response Critics. *The Washington Post* September 6, 2005. http://www.washingtonpost.com/wp-dyn/conmtent/artcile/2005/09/06/AR2005090501590_p. (Accessed 25.08.2018).
222. Banfield, E. C. *"Ends and Means in Planning" in a Reader in Planning Theory*; Pergamon Press: New York, 1973.
223. Wilson, T. D. Strangers to Ourselves: Discovering the Adaptive Unconscious; The Belknap Press of Harvard University Press: Cambridge, Massachusetts, and London, England, 2012; pp. 31–33, 211, 212.
224. Taleb, N. N. *The Black Swan: The Impact of the Highly Improbable*; Random House: New York, 2007.
225. Finkelstein, S.; Whitehead, J., Campbell, A. *Boston.* Harvard Business Press: Massachusetts, 2008; p.xii.
226. Thaler, R. H. *Improving Decision Making about Health, Wealth, and Happiness*; Yale University Press: New Haven, CT, 2008.
227. Haider, Ovid. Metamorphoses, Trans *David Raeburn*. Penguin: London, 2003; p. 32.
228. Schwartz, J. Big Storms Get Worse; *And We're Not Helping. The New York Times September* **2018**, *21*, p. 18. http://emergencymgmnt.com/disaster/Emergency-Management-Climate-Change.html.
229. Klein, K. R.; Nagel, N. E. *Disaster Mag Response* **2007**; *5*(2), 56–61. https://www.ncbi.nlm.nih.gov.pubmed/17517364 (Accessed 26.07.2018).

230. Lehman, R. J.; Semelsberger, D. *Take a Load Off Fannie: The GSES and Uninsured Earthquake Risk. RStreet: Free Market Solutions.* https://mail.yahoo.com/neo/ie_blank (Accessed 20.09.2018).

231. Erdman, J. *Heat Records Shattered in German, France; The Netherlands in June/July, 2015 European Heat Wave;* The Weather Channel (weather.com).

232. Richards, E. P. Climate Change Law and Policy Project. https://us-mg204.mail.yahoo.com/neo/launch?.partner=sbc&.rand=0ool53o5qbmcc (Accessed 20.11.2018).

233. Centers for Disease Control and Prevention (CDC). http://www.bt.cdc.gov/masscasualties/research/community.asp. (Accessed 23.06.2013).

234. Coppola, D. *Introduction to International Disaster Management.* Elsevier: New York and Oxford, 2015; p. xxiii.

235. Jacobini, H. B. *International Law: A Text.* Dorsey Press and Die Limited: Wales, Wisconsin, 1968.

236. Anderson, M.; Woodrow, P. *Rising from the Ashes: Development Strategies in Times of Disaster.* Westview Press: Boulder, CO, 1989, p. 19.

237. Edmond, K. *NGOs and the Business of Poverty in Haiti North American Congress on Latin America (NACLA),* April 5, 2010. https://nacla.org/news/ngos-and-business-poverty-haiti (Accessed 13.03.2019).

238. Philipps, B. D.; Neal, D. M.; Webb, G.R *Introduction to Emergency Management.* CRC Press: Boca Raton, London & New York, 2012. pp. 67, 68, 71, 272, 273, 415–420.

239. Dimento, J. F. C.; Doughman, P., Eds.; Climate Change: How the World is Responding, 2007. In *Climate Change; What is Means for Us, Our Children and Our Grandchildren.* MIT Press: p. 125.

240. Rabe, B. *Statehouse and Greenhouse: The Emerging Politics of American Climate Change Policy.* Pew Center on Global Climate Change: Washington, DC, 2002; p. 22.

241. Kahn, M.; Climatopolis, E. *How Our Cities Will Thrive in the Hotter Future;* Basic Books: New York, 2019; pp. 17–46.

242. "Chicago's Recovery." New York Times, October 23, 1881. http://query.nytimes.com/mem/archivefree/pdf?res=9A01E5DD103EE433A25751C1A9669D9609FDCF (Accessed 14.03.2010).

243. "Federal Coordinator for Gulf Coast Rebuilding Douglas O'Dell Hosts Federal Inspectors General Strategy Meeting". Press release, Department of Homeland Security, May 1, 2010. http://www.dhs.gov/xnews/releases/pr_1210791829291.shtm (Accessed 14.03.2010).

244. Dimento, J. F. C.; Doughman, P., Eds. *Climate Change: What is Means for Us, Our Children, and our Grandchildren;* MIT Press: Cambridge, Massachusetts & London, England, 2017; p. viii.

245. Cuny, F. *Disasters and Development;* Doubleday: NY, 1983.

246. Barton, A. *Communities in Disasters;* Intertech Press: Dallas, TX, 1970.

247. Sherlock, M. F.; Ravelle, J. G. *An Overview of the Nonprofit and Charitable Sector;* Congressional Research Service: Washington, DC, 2009; pp. 3, 16.

248. Stauffer, A.; Ford, C. The PEW Charitable Trust: Article, September 6, 2017. In *Natural Disaster Lend Each Other a Hand: Nationwide Mutual Assistance Compact Promises Aid From Neighbors.* Available at: https://www.pewtrusts.org/en/research-and-analysis/articles/2017/09/06/in-natural-disasters-states-lend-each-other-a-hand.

249. Phillips-Fein, K. The Fight over Big Government was Bitter from the Start. *The Atlantic* **2019**, 322(2), 32–34. Ms. Phillips-Fein was generally citing the Winter War, by Eric Rauchway, New York, Basic Books, 2018.

250. Kahneman, D.; Klein, G. Conditions for Intuitive Expertise: A Failure to Disagree. *American Psychologist* **2009**, 64(6), 505-26.

Index

Note: Page numbers followed by *f* indicate figures.

Printed in the United States
By Bookmasters